Communications in Computer and Information Science 1505

Yunjun Gao · An Liu · Xiaohui Tao ·
Junying Chen (Eds.)

Web and Big Data

APWeb-WAIM 2021 International Workshops

KGMA 2021, SemiBDMA 2021, DeepLUDA 2021
Guangzhou, China, August 23–25, 2021
Revised Selected Papers

Editors
Yunjun Gao
Zhejiang University
Hangzhou, China

An Liu ⓘ
Soochow University
Suzhou, China

Xiaohui Tao ⓘ
University of Southern Queensland
Toowoomba, QLD, Australia

Junying Chen ⓘ
South China University of Technology
Guangzhou, China

ISSN 1865-0929 ISSN 1865-0937 (electronic)
Communications in Computer and Information Science
ISBN 978-981-16-8142-4 ISBN 978-981-16-8143-1 (eBook)
https://doi.org/10.1007/978-981-16-8143-1

This Springer imprint is published by the registered company Springer Nature Singapore Pte Ltd.
The registered company address is: 152 Beach Road, #21-01/04 Gateway East, Singapore 189721, Singapore

Preface

The Asia Pacific Web (APWeb) and Web-Age Information Management (WAIM) Joint International Conference on Web and Big Data (APWeb-WAIM) is a leading international conference for researchers, practitioners, developers, and users to share and exchange their cutting-edge ideas, results, experiences, techniques, and applications in connection with all aspects of web and big data management. The conference invites original research papers on the theory, design, and implementation of data management systems. As the 5th edition in the increasingly popular series, APWeb-WAIM 2021 was held as a virtual conference in Guangzhou, China, during 23–25 August, 2021, attracting more than 600 participants from all over the world.

Along with the main conference, the APWeb-WAIM workshops provide an international forum for researchers to discuss and share their pioneered works. The APWeb-WAIM 2021 workshop volume contains a tutorial summary and the papers accepted for three workshops held in conjunction with APWeb-WAIM 2021. The three workshops were selected after a public call-for-proposal process, and each of them had a focus on a specific area that contributed to the main themes of the APWeb-WAIM conference. After the single-blinded review process, out of 28 submissions, the three workshops accepted a total of 11 full papers, marking an acceptance rate of 39.29%. The three workshops were as follows:

- The Fourth International Workshop on Knowledge Graph Management and Applications (KGMA 2021)
- The Third International Workshop on Semi-structured Big Data Management and Applications (SemiBDMA 2021)
- The Second International Workshop on Deep Learning in Large-scale Unstructured Data Analytics (DeepLUDA 2021)

As a joint effort, all organizers of APWeb-WAIM conferences and workshops, including this and previous editions, have made not only their work a proud but also APWeb-WAIM a valuable trademark. We would like to express our thanks to all the workshop organizers and Program Committee members for their great efforts in making the APWeb-WAIM 2021 workshops such a great success. Last but not least, we are grateful to the main conference organizers for their leadership and generous support, without which this APWeb-WAIM 2021 workshop volume wouldn't have been possible.

October 2021

Yunjun Gao
An Liu
Xiaohui Tao

Organization

APWeb-WAIM 2021 Workshop Co-chairs

Yunjun Gao   Zhejiang University, China
An Liu   Soochow University, China
Xiaohui Tao   University of Southern Queensland, Australia

APWeb-WAIM 2021 Publication Chair

Junying Chen   South China University of Technology, China

KGMA 2021

Workshop Chairs

Qingpeng Zhang   City University of Hong Kong, China
Xin Wang   Tianjin University, China

Program Committee

Huajun Chen   Zhejiang University, China
Wei Hu   Nanjing University, China
Saiful Islam   Griffith University, Australia
Jiaheng Lu   University of Helsinki, Finland
Jianxin Li   Deakin University, Australia
Ronghua Li   Beijing Institute of Technology, China
Jeff Z. Pan   University of Aberdeen, UK
Jijun Tang   University of South Carolina, USA
Haofen Wang   Tongji University, China
Hongzhi Wang   Harbin Institute of Technology, China
Junhu Wang   Griffith University, Australia
Meng Wang   Southeast University, China
Xiaoling Wang   East China Normal University, China
Xuguang Ren   Inception Institute of Artificial Intelligence, UAE
Guohui Xiao   Free University of Bozen-Bolzano, Italy
Zhuoming Xu   Hohai University, China
Qingpeng Zhang   City University of Hong Kong, China
Xiaowang Zhang   Tianjin University, China

SemiBDMA 2021

Workshop Chairs

Ge Yu	Northeastern University, China
Baoyan Song	Liaoning University, China
Xiaoguang Li	Liaoning University, China
Linlin Ding	Liaoning University, China
Yuefeng Du	Liaoning University, China

Program Committee

Bo Ning	Dalian Maritime University, China
Yongjiao Sun	Northeastern University, China
Yulei Fan	Zhejiang University of Technology, China
Guohui Ding	Shenyang Aerospace University, China
Bo Lu	Dalian Nationalities University, China
Jiajia Li	Shenyang Aerospace University, China
Rui Zhu	Shenyang Aerospace University, China
Yue Zhao	Shenyang University, China

DeepLUDA 2021

Organizers

Tae-Sun Chung	Ajou University, South Korea
Jianming Wang	Tiangong University, China
Zhetao Li	Xiangtan University, China

Workshop Chair

Rize Jin	Tiangong University, China

Program Committee

Liangfu Lu	Tianjin University, China
Ziyang Liu	Kyonggi University, South Korea
Yunbo Rao	University of Electronic Science and Technology of China, China
Se Jin Kwon	Kangwon National University, South Korea
Caie Xu	Zhejiang University of Science and Technology, China
He Li	Xidian University, China
Yenewondim Biadgie	Ajou University, South Korea
Muhammad Attique	Sejong University, South Korea
Zhen Wang	Tianjin University of Finance and Economics, China

Huayan Zhang   Tiangong University, China
Gaoyang Shan   Ajou University, South Korea
Xun Luo   Tianjin University of Technology, China
Joon-Young Paik   Tiangong University, China
Guanghao Jin   Beijing Polytechnic University, China
Jing Yang   Guangdong University of Foreign Studies, China
Yanji Piao   Yanbian University, China
Zhen Zhang   Tiangong University, China

Industrial Big Data Cleaning (Tutorial)

Hongzhi Wang and Xiaoou Ding

Harbin Institute of Technology, China

Abstract. Nowadays, it is realized that industrial data is a kind of valuable intangible assets. At the meanwhile, the data collected from industrial field is faced with various quality problems. Practicable data cleaning techniques are urgently needed in industrial applications. In this tutorial, we will synthesize and survey the research and development in industrial data cleaning, including industrial data cleaning workflow, error detection and error repairing in industrial data, knowledge modelling, the state-of-art data cleaning tools and systems, and the new challenges and opportunities brought by industrial applications.

Contents

The Second International Workshop on Deep Learning in Large-scale Unstructured Data Analytics

The Fourth International Workshop on Knowledge Graph Management and Applications

Should I Stay or Should I Go: Predicting Changes in Cluster Membership

Evangelia Tsoukanara[1(✉)], Georgia Koloniari[1], and Evaggelia Pitoura[2]

[1] Department of Applied Informatics, University of Macedonia, Thessaloniki, Greece
{etsoukanara,gkoloniari}@uom.edu.gr
[2] Computer Science and Engineering, University of Ioannina, Ioannina, Greece
pitoura@cse.uoi.gr

Abstract. Most research on predicting community evolution focuses on changes in the states of communities. Instead, we focus on individual nodes and define the novel problem of predicting whether a specific node stays in the same cluster, moves to another cluster or drops out of the network. We explore variations of the problem and propose appropriate classification features based on local and global node measures. Motivated by the prevalence of machine learning approaches based on embeddings, we also introduce efficiently computed distance-based features using appropriate node embeddings. In addition, we consider chains of features to capture the history of the nodes. Our experimental results depict the complexity of the different formulations of the problem and the suitability of the selected features and chain lengths.

Keywords: Cluster evolution · Embeddings · Feature selection · Classification

1 Introduction

For the problem of community evolution, communities are monitored across time and their properties studied to attain useful conclusions. Such conclusions can then be exploited so as to predict community changes through time. Most works model community evolution through a predefined set of events, such as community growth or shrinkage, community merging or splitting and so on [5, 7,12,16]. The problem is then modeled as a classification problem, where given a community history, the next event in its evolution is predicted.

However, in many applications, it is important not only to predict the behavior of a community but also of individual community members. Let us consider customers that are connected via the common products they buy. These customers can be clustered according to the companies and products they prefer. Focusing on individuals in such clusters, makes it possible for companies to identify and reward loyal customers (community members) or take preemptive measures to change the behavior of the ones that seem less dedicated.

© Springer Nature Singapore Pte Ltd. 2021
Y. Gao et al. (Eds.): APWeb-WAIM 2021 Workshops, CCIS 1505, pp. 3–15, 2021.
https://doi.org/10.1007/978-981-16-8143-1_1

To this end, we introduce a novel community evolution prediction problem defined at node level. For each individual node, there are three possible events: the node may (a) stay in the same cluster, (b) move to a different cluster, or (c) drop out of the network. We study variations of the problem by considering combinations and subsets of the possible evolution events, and evaluate the use of different classification features for solving them. Firstly, we consider classic features based on popular node measures. Motivated by the advances of machine learning methods for community detection that use node embeddings, we also propose features based on distances between such embeddings. In particular, we deploy the ComE [4] approach that provides both node and community embeddings. Features of both methods, defined at cluster and out of cluster or network level, are combined into chains, modeling the evolution history of the nodes through time [16]. Our experimental results show that the problems we defined are not trivial, and that the proposed distance-based embeddings perform almost as well as the classic ones, while being computed much more efficiently.

The rest of the paper is structured as follows. Section 2 briefly describes related work. In Sect. 3, we formulate the novel problem and its variations. Section 4 presents our classification features and their modeling into chains. Section 5 includes our experimental results, while Sect. 6 concludes.

2   Related Work

We discuss related research, first, on community evolution and then, on node and graph embeddings.

Community Evolution. After discovering communities in evolving networks [15], their properties are studied by mapping corresponding communities through successive network snapshots [1]. Community evolution is assessed with measures such as its growth and disappearance rate [20], or its life expectancy [10].

For predicting community evolution [5,12,16], events such as community growth, shrinkage, merging and splitting are defined. The problem is modeled as a classification problem in which features based on the structural properties of communities are exploited, and given the history of a community the next event in the community's evolution is predicted. In our work, instead of communities, we focus on nodes, and introduce a new problem aiming at predicting changes in the nodes' cluster membership through time.

Node Embeddings. An embedding is the transformation of a high-dimensional space to a low-dimension vector. For graphs, the focus is mostly on node embeddings that preserve network structure. Deepwalk [14] learns node embeddings that capture second-order proximity (i.e., proximity between shared neighbors) by simulating short random walks and applying the Skip-gram algorithm. LINE [17] employs edge-sampling, and to preserve both first-order (i.e., ties between neighbors) and second-order proximity, it is first trained separately and then the two resulting embeddings are concatenated. Node2vec [6] extends Deepwalk by generating biased random walks to explore diverse node neighborhoods. Finally, GraRep [3] and HOPE [9] derive embeddings that capture high-order proximity.

Besides, link prediction, node classification and visualization, node and graph embeddings are also used for community detection [8, 19]. ComE (Community Embedding) [4] is a framework that jointly solves both community detection, and learning node and community embeddings. The intuition is that node embeddings that capture community-aware proximity, can assist community detection, while community embeddings can in turn improve node embeddings. In our work, we deploy ComE and explore whether node and community embeddings can be used to derive predictive features.

3 Problem Formulation

A social network is often represented as a graph $G = (V, E)$, where V is the set of nodes (vertices) and E is the set of edges. A temporal social network is a network that changes over time and is represented as a sequence of graphs $\{G_1, G_2, \ldots, G_n\}$, where $G_i = (V_i, E_i)$, represents a snapshot of graph G at time i, and V_i and E_i are the node and edge sets at time i respectively. Let $\mathcal{C}_i = \{C_i^1, C_i^2, \ldots, C_i^m\}$ be a clustering of G_i consisting of m clusters, such that $C_i^j \cap C_i^k = \emptyset, 1 \leq j, k \leq m, j \neq k$. For two clusterings $\mathcal{C}_i$ and $\mathcal{C}_{i+1}$ at consecutive timeframes i and $i + 1$, we assume cluster C_{i+1}^j is the evolution of cluster C_i^j. Similarly, for multiple consecutive timeframes, $C_1^j, C_2^j, \ldots, C_n^j$ denotes the evolution of cluster C^j in time period $[1, n]$.

Inspired by the idea of community evolution, we study the problem at node level by aiming to predict how node memberships in clusters evolve through time. We discern between different states that a node can have with respect to its cluster membership in the next timeframe, i.e., a node can stay in the same cluster, move to another or drop out of the network. Based on the above, we define our problem as follows:

Definition 1 (Stay\Move\Drop ($\mathcal{SMD}$) Problem). *Given a sequence of clusterings $\mathcal{C}_1, \ldots, \mathcal{C}_i$, corresponding to a consecutive set of timeframes for a graph G, and node $v \in C_i^j$, predict the state of node v regarding its evolution in the next timeframe $i + 1$ as state:*

- *stay, $\mathcal{S}$: node v stays at the same cluster in $i + 1$, that is $v \in C_{i+1}^j$,*
- *move, $\mathcal{M}$: node v moves to another cluster in $i + 1$, that is $v \in C_{i+1}^k$, $k \neq j$, and*
- *drop, $\mathcal{D}$: node v drops out of the network, that is $v \notin V_{i+1}$.*

Thus, we define a classification problem with 3 classes, labeled *stay*, *move* and *drop*. Given the history of a node in a given time period, defined as a sequence of distinct timeframes, we predict its class in the next time frame.

If we are only interested in discerning between loyal cluster members and members likely to leave, we may simplify our problem to a binary classification problem. This first alternative problem, is derived by merging states *move* and *drop*, in one class *leave*. Thus, the classes are reformed as follows:

Stay\Leave ($\mathcal{SL}$) Problem: the possible node events are reformed as state:

- *stay, $\mathcal{S}$*: node v stays at the same cluster in $i+1$, that is $v \in C_{i+1}^j$, and
- *leave, $\mathcal{L}$*: node v does not remain in the same cluster, that is $v \notin C_{i+1}^j$.

Finally, we omit the third class of the $\mathcal{SMD}$ problem, and only provide predictions for nodes that remain in the network in timeframe $i+1$. For the third variation, we have:

Stay\Move ($\mathcal{SM}$) Problem.

- *stay, $\mathcal{S}$*: node v stays at the same cluster in $i+1$, that is $v \in C_{i+1}^j$, and
- *move, $\mathcal{M}$*: node v moves to another cluster in $i+1$, that is $v \in C_{i+1}^k$, $k \neq j$.

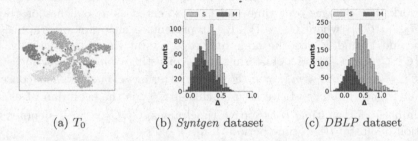

(a) T_0 (b) *Syntgen* dataset   (c) *DBLP* dataset

Fig. 1. (a) ComE clustering and Δ distribution for (b) *Syntgen* and (c) *DBLP*.

4   Predictive Features and Historical Chains

To solve the classification problems we define, the evaluation and selection of appropriate predictive features is required. We discern between two basic types of features, based on structural network measures that are usually exploited for community evolution predictions, and on distances between embeddings that we propose. As all three problems we introduce are variations of the same classification problem, we define and evaluate the same features for all problems.

Classic Features. Classic features, defined on node level, provide information about the structural role of a node in the network. In particular, we select: (a) *degree* measuring the connections of a node, (b) *betweenness*, measuring the number of shortest paths that pass through a node, (c) *closeness*, that measures the distance of a node to all other nodes, and (d) *eigenvector* centrality measuring the influence of a node in the network, defined on cluster and network level.

Aggregated at community level, classic features offer insights for predicting community evolution [12,16], while for individual nodes, such measures also contain information about their evolution tendencies. For instance, an influential central node with high degree is less likely to drop out of the network compared to a low-degree remote node. Similarly focusing on community structure, well-connected core nodes within a community are more likely to stay in their cluster compared to loosely connected border nodes. Thus, we also differentiate between features defined at cluster level (*in*) and at network level (*out*).

Embeddings-Based Features. Since node embeddings are low dimensional vector representations of nodes that capture structural graph properties, we propose defining predictive features based on such embeddings. The ComE [4] framework provides an appropriate solution by solving both community detection and embeddings learning jointly, focusing on deriving embeddings that capture community-aware proximity between nodes. ComE uses as input a graph's edge list and the number of target clusters k and outputs: (i) for each node its embedding along with its community membership and (ii) for each community, which is defined as a multivariate Gaussian distribution, its embedding parameters, that is, a median vector (i.e., the embedding of the mean of the community) and a covariance matrix. Figure 1a depicts the node embeddings for the six clusters detected by ComE for a synthetic dataset generated by the Syntgen [13] generator for timeframe T_0, where we notice that ComE manages to detect reasonable well-separated clusters with similar node embeddings as projected in 2-d.

We define features on cluster and network level, exploiting the outputs of ComE. For computing network level features, we exclude the nodes of the cluster the given node belongs to, to avoid cases of identical network and cluster features. Let ϕ_v be the embedding of node v, and ϕ_C the embedding of the median of community C. Without loss of generality we define our features using the euclidean distance (d) between pairs of embeddings. Cosine and $L1$ distances were also evaluated, but euclidean were used as they performed slightly better. In particular, for a node v such that $v \in C$, we define the following features:

Cluster level:

- distance from cluster median: $d(\phi_v, \phi_C)$
- distance from least similar cluster member: $\max_{\forall u \in C} d(\phi_v, \phi_u)$
- distance from most similar cluster member: $\min_{\forall u \in C} d(\phi_v, \phi_u)$
- avg. distance from all cluster members: $avg_{\forall u \in C} d(\phi_v, \phi_u)$

Network level:

- min. distance from other cluster median: $\min_{\forall C' \neq C} d(\phi_v, \phi_{C'})$
- distance from least similar node out of the cluster: $\max_{\forall u \notin C} d(\phi_v, \phi_u)$
- distance from most similar node out of the cluster: $\min_{\forall u \notin C} d(\phi_v, \phi_u)$
- avg. distance from all nodes not in the cluster: $avg_{\forall u \notin C} d(\phi_v, \phi_u)$

Our approach is based on the idea that the embeddings of nodes in a cluster are more similar. Thus, if a node has an embedding that is similar to nodes of other clusters, it is more likely to change or leave its cluster. In Fig. 1b and Fig. 1c, we plot the distribution of the difference, Δ, between $\min_{\forall C' \neq C} d(\phi_v, \phi_{C'})$ and $d(\phi_v, \phi_C)$ features, for classes *stay* and *move* for a synthetic dataset generated by Syntgen and a citation DBLP dataset based on [18]. In both figures, we notice that nodes that move to another cluster tend to have lower or even negative Δ compared to the ones that remain in the same cluster. About 60% of the *move* nodes have Δ less than 0.20 for both datasets, while more than 60% of the *stay* nodes have Δ higher than 0.28. Therefore, the values of the various distance measures can provide insight on the properties and behavior of a node, and thus, are appropriate as predictive features for our classification problems.

Historical Chains of Features. The evolution of a community is tracked in successive network snapshots that correspond to successive timeframes. Thus, all features we describe can be measured for each timeframe. Let us assume a set of k predictive features for each node v, and a time period $[1, \ldots, n]$, where $f_i^j(v)$ denotes the j-th feature of node v at time i. Further, for every pair of consecutive timeframes i and $i + 1$ in the given time period, the state (label), $l_{i+1}(v)$, of node v can be recorded.

To model the history of the node and exploit it to derive more accurate predictions, we utilize historical chains of features as defined in [16]. In particular, we have as final features for v: $\{f_1^1(v), \ldots, f_1^k(v)\}$, $l_2(v)$, $\{f_2^1(v), \ldots, f_2^k(v)\}$, $l_3(v)$ $\ldots$, $\{f_n^1(v), \ldots, f_n^k(v)\}$, while $l_{n+1}(v)$ is the label to be predicted.

Though we have defined here one chain to model the entire node history, the use of subchains of various lengths can also be deployed. While a longer history will provide more information regarding the history of a node, it would limit the number of nodes for which a prediction can be made.

Table 1. Snapshot structure

Sn#	DBLP				Email-eu				Syntgen			
	Nodes	Edges	$|\mathcal{C}|$	$\mathcal{Q}$	Nodes	Edges	$|\mathcal{C}|$	$\mathcal{Q}$	Nodes	Edges	$|\mathcal{C}|$	$\mathcal{Q}$
0	14731	120192	17	0.681	750	4740	11	0.515	1583	8955	6	0.525
1	16801	143404	16	0.676	745	5077	11	0.431	1443	7936	5	0.480
2	17756	156393	16	0.649	742	4578	10	0.528	1367	7249	5	0.459
3	14765	120370	17	0.659	742	4886	9	0.487	1567	8154	5	0.476
4	10898	69005	16	0.660	749	5072	11	0.388	1575	7993	5	0.483
5					739	4819	10	0.521	1402	6978	5	0.465
6					759	4846	10	0.521	1526	7629	5	0.472
7					808	5405	11	0.410	1416	6973	5	0.462
8					772	4880	13	0.391	1409	6853	5	0.470
9					785	5169	12	0.533	1594	7794	6	0.490

5   Evaluation

We experimentally study all three classification problems while comparing different sets of predictive features.

5.1   Datasets

We use three datasets for our evaluation, two real and one synthetic.

DBLP: *DBLP* is a citation network[1] that includes additional information about the publications, such as year of publication and fields of study they belong to

---

[1] https://www.aminer.org/citation.

Table 2. Performance for the $\mathcal{SM}$ problem

Data	Feat.	Class	ComE				Classic			
			P	R	F1	Acc	P	R	F1	Acc
DBLP	in	$\mathcal{S}$	0.821	0.940	0.876	0.804	0.837	0.941	0.886	0.821
		$\mathcal{M}$	0.715	0.425	0.533		0.746	0.485	0.587	
	out	$\mathcal{S}$	0.800	0.942	0.865	0.784	0.783	0.927	0.849	0.757
		$\mathcal{M}$	0.679	0.339	0.452		0.577	0.279	0.376	
	all	$\mathcal{S}$	0.836	0.954	**0.891**	0.828	0.839	0.948	**0.890**	0.827
		$\mathcal{M}$	0.787	0.476	**0.592**		0.769	0.490	**0.599**	
Email-eu	in	$\mathcal{S}$	0.807	0.926	0.862	0.796	0.797	0.912	0.850	0.778
		$\mathcal{M}$	0.757	0.507	0.606		0.712	0.483	**0.574**	
	out	$\mathcal{S}$	0.778	0.933	0.848	0.770	0.732	0.910	0.812	0.709
		$\mathcal{M}$	0.733	0.409	0.524		0.569	0.261	0.357	
	all	$\mathcal{S}$	0.823	0.934	**0.875**	0.816	0.789	0.925	**0.852**	0.778
		$\mathcal{M}$	0.792	0.554	**0.652**		0.731	0.452	0.558	
Syntgen	in	$\mathcal{S}$	0.675	0.692	0.683	0.668	0.707	0.719	**0.713**	0.700
		$\mathcal{M}$	0.661	0.643	0.652		0.693	0.680	0.687	
	out	$\mathcal{S}$	0.638	0.704	0.670	0.640	0.642	0.669	0.655	0.635
		$\mathcal{M}$	0.643	0.572	0.606		0.628	0.599	0.613	
	all	$\mathcal{S}$	0.693	0.725	**0.708**	0.691	0.710	0.713	0.711	0.701
		$\mathcal{M}$	0.690	0.656	**0.672**		0.691	0.687	**0.689**	

Table 3. Performance for the $\mathcal{SL}$ problem

Data	Feat.	Class	ComE				Classic			
			P	R	F1	Acc	P	R	F1	Acc
DBLP	in	$\mathcal{S}$	0.626	0.676	0.650	0.922	0.715	0.687	0.700	0.937
		$\mathcal{L}$	0.961	0.951	0.956		0.962	0.967	0.965	
	out	$\mathcal{S}$	0.632	0.664	0.647	0.922	0.680	0.648	0.664	0.929
		$\mathcal{L}$	0.959	0.953	0.956		0.958	0.963	0.960	
	all	$\mathcal{S}$	0.680	0.710	**0.695**	0.933	0.728	0.690	**0.709**	0.939
		$\mathcal{L}$	0.965	0.960	**0.962**		0.963	0.969	**0.966**	
Email-eu	in	$\mathcal{S}$	0.796	0.923	0.855	0.820	0.799	0.892	**0.843**	0.809
		$\mathcal{L}$	0.869	0.680	0.762		0.827	0.698	**0.756**	
	out	$\mathcal{S}$	0.771	0.915	0.837	0.795	0.725	0.882	0.796	0.740
		$\mathcal{L}$	0.847	0.633	0.724		0.775	0.548	0.642	
	all	$\mathcal{S}$	0.809	0.934	**0.867**	0.836	0.793	0.892	0.840	0.804
		$\mathcal{L}$	0.889	0.703	**0.784**		0.826	0.686	0.749	
Syntgen	in	$\mathcal{S}$	0.656	0.654	0.655	0.708	0.696	0.690	**0.693**	0.741
		$\mathcal{L}$	0.747	0.748	0.747		0.774	0.778	**0.776**	
	out	$\mathcal{S}$	0.634	0.660	0.646	0.694	0.631	0.625	0.628	0.686
		$\mathcal{L}$	0.742	0.720	0.731		0.726	0.731	0.729	
	all	$\mathcal{S}$	0.670	0.682	**0.676**	0.723	0.693	0.684	0.688	0.738
		$\mathcal{L}$	0.763	0.753	**0.758**		0.770	0.778	0.774	

[18]. Similarly to [4], we filter papers with primary of study as NLP, Databases, Networking, Data Mining and Computer Vision. We construct five snapshots, for years 2015 to 2019, by maintaining publications of the given year and adding cited papers that belong to the selected fields regardless of their publication year. To attain a denser network, we sample nodes with degree ≥ 20 and build the induced undirected subgraph.

Email-eu: The *Email-eu*[2] [11] dataset consists of incoming and outgoing e-mails between members of a large European institution in a period of 803 days. We consider the network undirected and split the data into 10 balanced snapshots.

Syntgen: To further investigate the impact of network properties on our problems, we use the Syntgen generator [13] that creates temporal undirected networks simulating real networks using explicit specifications, like degree distributions and cluster sizes, as well as implicitly controlling the perseverance of nodes popularity over time. The intra-cluster to total degree ratio determines cluster density. A high ratio leads to dense well-separated communities, while low values exhibit no clustering. We set the default ratio to 0.7.

We apply ComE [4] to detect communities at each snapshot. To determine an appropriate number of clusters k as input for ComE, for *DBLP* and *Email-eu*, we first apply the Louvain [2] community detection method that selects the k that maximizes modularity. As to *Syntgen*, we use as k the number of clusters obtained from the generator. Clusters are mapped across different snapshots based on the majority of their common nodes. Table 1 presents the number of nodes and edges as well as the number of clusters ($|\mathcal{C}|$) and modularity (Q) for each network snapshot for all datasets.

With regards to historical chains, *DBLP* with 5 snapshots can form 2-length up to 4-length chains, while *Email-eu* and *Syntgen* with 10 snapshots can form from 2-length up to 9-length chains. Trying to balance between more information that longer chains provide and the ability to provide predictions for more nodes, we use as default chain length 5 for both *Email-eu* and *Syntgen*, and 2 for *DBLP*.

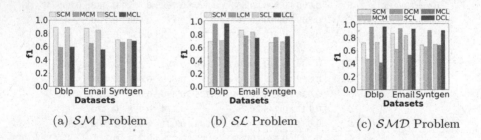

(a) $\mathcal{SM}$ Problem   (b) $\mathcal{SL}$ Problem   (c) $\mathcal{SMD}$ Problem

Fig. 2. Macro f1 score per class.

---

[2] https://snap.stanford.edu/data/email-Eu-core-temporal.html.

5.2   Experimental Results

We report results using the Random Forest classifier, which performed better compared to other classifiers we tried (e.g. Logistic Regression, Naive Bayes). We apply stratified 5-fold cross-validation to preserve the same class distribution in both train and test sets. Tables 2, 3 and 4 illustrate precision, recall and f1-score per class, and accuracy for each problem when using ComE and Classic features at cluster (in) and network level (out), and their combination (all) for each dataset. We denote with bold the best f1-score for each class per dataset and problem. As structural features are typically used in community evolution prediction, we consider Classic features as a baseline method to compare the proposed ComE-based features.

General Observations. A first observation derived by Table 2, Table 3 and Table 4 is that while the three classification problems are not trivial, both Classic and ComE based features show promising initial results, achieving high accuracy

Table 4. Performance for the SMD problem

Data	Feat.	Class	ComE				Classic			
			P	R	F1	Acc	P	R	F1	Acc
DBLP	in	S	0.589	0.818	0.685	0.905	0.666	0.778	0.718	0.917
		M	0.538	0.358	0.430		0.587	0.315	0.410	
		D	0.973	0.941	0.957		0.962	0.962	0.962	
	out	S	0.587	0.821	0.685	0.904	0.627	0.744	0.680	0.911
		M	0.491	0.250	0.331		0.463	0.161	0.239	
		D	0.972	0.944	0.958		0.961	0.966	0.963	
	all	S	0.621	0.853	**0.718**	0.914	0.682	0.778	**0.727**	0.921
		M	0.601	0.384	**0.468**		0.616	0.310	**0.411**	
		D	0.976	0.945	**0.960**		0.963	0.967	**0.965**	
Email-eu	in	S	0.795	0.932	0.858	0.816	0.783	0.909	**0.841**	0.790
		M	0.748	0.517	0.611		0.653	0.468	**0.545**	
		D	1	0.881	0.936		0.996	0.881	**0.935**	
	out	S	0.762	0.925	0.836	0.785	0.718	0.897	0.798	0.732
		M	0.686	0.413	0.515		0.524	0.272	0.357	
		D	1	0.881	0.936		0.993	0.881	0.933	
	all	S	0.797	0.939	**0.862**	0.821	0.775	0.905	0.835	0.784
		M	0.765	0.520	**0.619**		0.647	0.451	0.530	
		D	1	0.881	**0.937**		0.990	0.881	0.932	
Syntgen	in	S	0.645	0.680	0.662	0.693	0.683	0.707	**0.695**	0.724
		M	0.634	0.646	0.640		0.668	0.696	0.681	
		D	1	0.828	0.906		1	0.828	0.905	
	out	S	0.620	0.692	0.654	0.680	0.620	0.658	0.638	0.670
		M	0.628	0.599	0.613		0.603	0.610	0.606	
		D	1	0.828	0.906		1	0.828	**0.906**	
	all	S	0.661	0.712	**0.685**	0.711	0.685	0.706	**0.695**	0.725
		M	0.659	0.657	**0.658**		0.668	0.698	**0.682**	
		D	1	0.828	**0.906**		1	0.828	**0.906**	

but showcasing that some classes are more difficult to predict than others. We also notice that *all* features perform better in most cases, and use them as our default features for the rest of our study.

$\mathcal{SM}$ **vs.** $\mathcal{SL}$ **vs.** $\mathcal{SMD}$. To compare the three problems, we illustrate the f1-score for each class for both types of features (ComE, denoted as CM, and Classic denoted as CL) for the three datasets in Fig. 2. For the $\mathcal{SM}$ problem, Fig. 2a shows that class *stay* performs better than *move* for *DBLP* and *Email-eu*. *DBLP* achieves the best performance with 0.891 for *stay* and 0.592 for *move* respectively for the ComE features and similar results for the Classic ones (Table 2). This is due to the imbalance in the real datasets between the two classes, with the majority of the nodes in class *stay*. In contrast, in the *Syntgen* dataset, classes are well-balanced and we notice similar performance for both. For the $\mathcal{SL}$ problem (Fig. 2b), we notice significant difference mainly on *DBLP*. In this case, class *stay* is underrepresented due to the network construction. Looking at Table 3, class *stay* achieves f1 0.695 and *leave* 0.962, for *DBLP* with feature type *all*. Finally, as we can see in Table 4 and Fig. 2c for the $\mathcal{SMD}$ problem, class *move* is heavily underrepresented in both *DBLP* and *Email-eu* resulting in a rather low f1-score. Summing up, the $\mathcal{SL}$ problem seems to have the best overall performance, while $\mathcal{SM}$ appears to have the worst. Apparently, class *move* is overall the most difficult to predict. Besides class imbalance that makes the problem more difficult, this occurs especially when the characteristics across communities are not significantly different, which is a similarity indicator between communities.

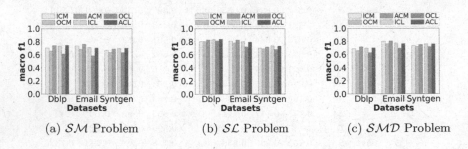

(a) $\mathcal{SM}$ Problem   (b) $\mathcal{SL}$ Problem   (c) $\mathcal{SMD}$ Problem

Fig. 3. Macro f1 score per feature category.

Selecting Appropriate Features. Next, we focus on the comparison between the different types of features *in* (I), *out* (O) and *all* (A) for ComE (CM) and Classic (CL) features and how they perform at each problem. As we can see in Fig. 3, *in* and *all* outperform *out* features. For ComE features *all* is the best choice for all problems, while for Classic features *in* performs sometimes better. This seems to depend on the dataset and not the problem we study, as we see that for *Email-eu* Classic *in* features perform better for all problems (with the exception of class *stay* on the $\mathcal{SM}$ problem). For the $\mathcal{SM}$ problem and ComE features, we notice a significant difference between *out* and *all* features (Fig. 3a). Macro f1 is 0.613 for *out* and 0.744 for *all* for the *DBLP* dataset and 0.584 for *out* and 0.704 for *all* for the *Email-eu* dataset. Overall, we do not observe significant differences between ComE and Classic features, deducing that the ComE features offer satisfying performance while being more efficiently computed. For instance, for the *DBLP* dataset, the computation time is 5493 s for ComE features, while Classic features are slower by an order, requiring 34618 s.

Influence of the Length of Historical Chains. Depicted in Fig. 4a, 4b, and 4c, we explore the effect of chains of features with varying length for the *DBLP* (D), *Email-eu* (E) and *Syntgen* (S). Both *Email-eu* and *Syntgen* show that f1 generally improves as chain length grows for all problems. The *Syntgen* dataset follows the same pattern for the $\mathcal{SL}$ and $\mathcal{SMD}$ problems, increasing and reaching its peak at chain length 8. For the *Email-eu*, we observe a temporary drop at length 7, while the highest score is reached at length 9, with 0.804 for the $\mathcal{SM}$ (Fig. 4a), 0.883 for the $\mathcal{SL}$ (Fig. 4b) and 0.850 for the $\mathcal{SMD}$ (Fig. 4c) problem respectively. As we have mentioned, while longer chains provide more information and more accurate predictions, they are not available for a large number of nodes. In particular, *Email-eu* consists of 6034 instances at 2-length chain and only 750 instances at its longest chain. Similarly, *Syntgen* consists of 11879 instances at 2-length and 1583 instances at 9-length chain. With regards to *DBLP*, the best score, 0.828, is achieved at the $\mathcal{SL}$ problem with chain length 2 with 0.828, but the history is too limited to derive safe comparative conclusions.

Influence of Intra-cluster to Total Degree Ratio. In the last experiment, we focus on the Syntgen generator producing different datasets with varying intra-cluster to total degree ratio, which determines the density within the constructed clusters compared to the overall network. In Fig. 4d, we notice a sharp increase on macro f1 for ratio 0.6 up to 0.8 for all problems. Lower ratio indicates poor clustering and thus is not suitable for our context. Beyond 0.8, behavior diverges. In such tightly connected communities, most nodes have similar roles in their community, making it difficult for a classifier to determine their behavior. As a conclusion, cases with ratio close to 0.5 that exhibit no locality, or close to 0.9 indicating almost disconnected communities, fail to achieve good results.

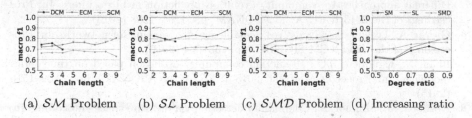

(a) $\mathcal{SM}$ Problem   (b) $\mathcal{SL}$ Problem   (c) $\mathcal{SMD}$ Problem   (d) Increasing ratio

Fig. 4. Macro f1 score (a), (b), (c) per chain length and (d) per degree ratio.

6 Conclusions

In this paper, we defined a novel problem, related to community evolution, that focuses on nodes and aims at predicting whether they will stay in their cluster, move to another or drop out of the network. We modeled the problem as a classification problem and with three variations. We determined appropriate features, based on both local and global node measures, and formed chains of features to take advantage of node history. We also proposed exploiting node and community embeddings derived by the ComE [4] framework to define distance based features. Our experimental results showed that the novel problem is not trivial, and the distance-based features performed similarly to the Classic ones, while requiring far less computation time. Next, we will consider alternative community learning approaches to derive node embeddings and define appropriate features.

Acknowledgements. Research work supported by the Hellenic Foundation for Research and Innovation (H.F.R.I.) under the "1st Call for H.F.R.I. Research Projects to Support Faculty Members & Researchers and Procure High-Value Research Equipment" (Project Number: HFRI-FM17-1873, GraphTempo).

References

1. Aynaud, T., Fleury, E., Guillaume, J.L., Wang, Q.: Communities in evolving networks: definitions, detection, and analysis techniques. In: Mukherjee, A., Choudhury, M., Peruani, F., Ganguly, N., Mitra, B. (eds.) Dynamics Onand Of Complex Networks. Volume 2: Applications to Time-Varying Dynamical Systems, pp. 159–200. Springer, New York (2013). https://doi.org/10.1007/978-1-4614-6729-8_9
2. Blondel, V.D., Guillaume, J.L., Lambiotte, R., Lefebvre, E.: Fast unfolding of communities in large networks. J. Stat. Mech. Theory Exp. **2008**(10), P10008 (2008)
3. Cao, S., Lu, W., Xu, Q.: GraRep: learning graph representations with global structural information. In: Proceedings of the 24th ACM CIKM, pp. 891–900 (2015)
4. Cavallari, S., Zheng, V.W., Cai, H., Chang, K.C.C., Cambria, E.: Learning community embedding with community detection and node embedding on graphs. In: Proceedings of the 2017 ACM CIKM, pp. 377–386 (2017)
5. Gliwa, B., Bródka, P., Zygmunt, A., Saganowski, S., Kazienko, P., Koźlak, J.: Different approaches to community evolution prediction in blogosphere. In: Proceedings of the 2013 IEEE/ACM ASONAM, pp. 1291–1298 (2013)
6. Grover, A., Leskovec, J.: Node2vec: scalable feature learning for networks. In: Proceedings of the 22nd ACM SIGKDD, pp. 855–864 (2016)
7. Ilhan, N., Öğüdücü, Ş.G.: Predicting community evolution based on time series modeling. In: Proceedings of the 2015 IEEE/ACM ASONAM, pp. 1509–1516 (2015)
8. Kozdoba, M., Mannor, S.: Community detection via measure space embedding. In: Advances in Neural Information Proceedings System 28, NIPS 2015, pp. 2890–2898 (2015)
9. Ou, M., Cui, P., Pei, J., Zhang, Z., Zhu, W.: Asymmetric transitivity preserving graph embedding. In: Proceedings of the 22nd ACM SIGKDD, pp. 1105–1114 (2016)
10. Palla, G., Barabási, A.L., Vicsek, T.: Quantifying social group evolution. Nature **446**, 664–667 (2007)
11. Paranjape, A., Benson, A.R., Leskovec, J.: Motifs in temporal networks. In: Proceedings of the Tenth ACM WSDM, pp. 601–610 (2017)

12. Pavlopoulou, M.E.G., Tzortzis, G., Vogiatzis, D., Paliouras, G.: Predicting the evolution of communities in social networks using structural and temporal features. In: 12th International Workshop on Semantic and Social Media Adaptation and Personalization, pp. 40–45 (2017)
13. Pereira, L.R., Lopes, R.J., Louçã, J.: Syntgen: a system to generate temporal networks with user-specified topology. J. Complex Netw. **4**, 1–26 (2019)
14. Perozzi, B., Al-Rfou, R., Skiena, S.: Deepwalk: online learning of social representations. In: Proceedings of the 20th ACM SIGKDD, pp. 701–710 (2014)
15. Rossetti, G., Cazabet, R.: Community discovery in dynamic networks: a survey. ACM Comput. Surv. **51**(2), 1–37 (2018)
16. Saganowski, S.: Predicting community evolution in social networks. In: Proceedings of the 2015 IEEE/ACM ASONAM, pp. 924–925 (2015)
17. Tang, J., Qu, M., Wang, M., Zhang, M., Yan, J., Mei, Q.: LINE: large-scale information network embedding. In: Proceedings of the 24th WWW, pp. 1067–1077 (2015)
18. Tang, J., Zhang, J., Yao, L., Li, J., Zhang, L., Su, Z.: ArnetMiner: extraction and mining of academic social networks. In: Proceedings of the 14th ACM SIGKDD, pp. 990–998 (2008)
19. Tian, F., Gao, B., Cui, Q., Chen, E., Liu, T.Y.: Learning deep representations for graph clustering. In: Proceedings of the 28th AAAI, pp. 1293–1299 (2014)
20. Toyoda, M., Kitsuregawa, M.: Extracting evolution of web communities from a series of web archives. In: Proceedings of the 14th ACM Conference on Hypertext and Hypermedia, pp. 28–37 (2003)

SMat-J: A Sparse Matrix-Based Join for SPARQL Query Processing

Ximin Sun[1(✉)], Ming Liu[1], Shuai Wang[2], Xiaoming Li[2], Bin Zheng[2], Dan Liu[2], and Hongshen Yu[3]

[1] State Grid Financial Technology Group, State Grid Electronic Commerce Co., Ltd., Beijing, China
liuming@sgec.sgcc.com.cn
[2] State Grid Ecommerce Technology Co., Ltd., Beijing, China
{wangshuai1,zhengbin}@sgec.sgcc.com.cn
[3] College of Intelligence and Computing, Tianjin University, Tianjin, China
hs_yu@tju.edu.cn

Abstract. In this demonstration, we present SMat-J, a SPARQL query engine for RDF datasets. It employs join optimization and data sparsity. We bifurcate SMat-J into three submodules e.g. Firstly, SM Storage (Sparse Matrix-based Storage) which lifts the storage efficiency, by storing valid edges, introduces a predicate-based hash index on the storage and generate a statistic file for optimization. Secondly, Query Planner which holds Query Parser and Query Optimizer. The Query Parser module parses a SPARQL query and transformed it into a query graph and the latter generates the optimal query plan based on statistical input from SM Storage. Thirdly, Query Executor module executes query in an efficient manner. Lastly, SMat-J evaluated by comparing with some well-known approaches like gStore and RDF3X on very large datasets (over 500 million triples). SMat-J is proved as significantly efficient and scalable.

Keywords: SPARQL · RDF · Sparse matrix · Join

1 Introduction

Resource Description Framework (RDF) [1] is a standard recommended by W3C. Over last decades, RDF has become the main standard for semantic data in the form of a triple: (subject, predicate, object). An RDF dataset can also be described, as a directed labeled graph, where, *subjects* and *objects* are vertices and triples are edges with labels (*predicate*). The queries over RDF dataset can be expressed, as SPARQL, which is officially recommended query language for RDF by W3C in 2008 and afterwords in 2013 its latest version SPARQL1.1 is circulated.

The existing Query engineer for SPARQL can be categorized into two major kinds in basis of storage strategies e.g. relation-based approach and graph-based

© Springer Nature Singapore Pte Ltd. 2021
Y. Gao et al. (Eds.): APWeb-WAIM 2021 Workshops, CCIS 1505, pp. 16–26, 2021.
https://doi.org/10.1007/978-981-16-8143-1_2

approach [2]. The existing stores take an RDF triple as a tuple in a ternary relation and the latter stores RDF data as a directed-labeled graph (e.g., gStore). The relation-based systems mainly apply a relational approach to store and index RDF data. Several systems like RDF-3X [9], TripleBit [10], BitMat [5], and BMatrix [11], employ specialized optimization techniques based on the features of RDF data and SPARQL queries. RDF-3X [9] build a set of indices that cover all possible permutations of S, P and O, in order to speed up the joins. TripleBit [10] uses a two dimension matrix to represent RDF triples, with subjects and objects as row and predicates as column. Then, it uses '1' to label the relation, otherwise uses '0'. Thus, there are only two '1' in each column, which are easy to be recorded. Furthermore, the triple matrix is divided into submatrices by the same properties and stored by column.

The graph-based approaches were proposed to store RDF triples in graph models, such as gStore [12], dipLODocus$_{[RDF]}$ [13], Turbo$_{HOM++}$ [14], and AMBER [15]. These approaches typically see SPARQL query processing as subgraph matching, which help reserve and query semantic information. gStore [12] maps all predicates and predicate values to binary bit strings which are then organized as a VS*-tree. Since every layer of VS*-tree is a summary of the whole RDF graph, gStore has capacity to process SPARQL query efficiently. dipLODocus [RDF] [13] starts by a mixed storage considering both graph structure of RDF data and requirement of data analysis, in order to find molecule clusters and help accelerate queries through clustering related data. AMBER [15] represents RDF data and SPARQL query as multigraphs while Turbo$_{HOM++}$ [14] transformes RDF graphs into normal data graphs.

All the above systems improve performance for query processing, however, they are computationally expensive in preprocessing due to their lack of parallelism. In addition, these systems are limited for large-scale datasets of real words such as DBpedia [3] and YAGO [4] with millions of triples. It is due to little consideration of an important feature called sparsity of those practical RDF data. The sparsity of RDF data means that the neighbors of each vertex in an RDF graph take a quite small proportion of the whole vertices. The sparsity of RDF data means that the neighbors of each vertex in an RDF graph take a quite small proportion of the whole vertices. In fact, the sparsity of RDF data exists everywhere. For an instance, There are over 99.41% nodes with at most 43° (sum of out-degrees and in-degrees) in DBpedia (42,966,066 nodes in total, Fig. 1(a)) and over 95.17% nodes with at most 39° in YAGO (38,734,252 nodes in total, Fig. 1(b)).

SPARQL is built on basic graph pattern (BGP) and SPARQL algebra operators. The join for concatenating variables is the core operation of SPARQL query evaluation, as BGP is the join of triple patterns. In an RDF graph, the product of their adjacent matrices can compute the join (concatenation) of two vertices. For instance, BitMat [5] employs matrices to compute join of SPARQL over RDF data. In proposing SMat-J framework, our contributions for this demonstration are as followed.

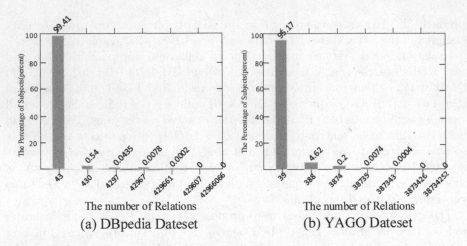

The number of Relations
(a) DBpedia Dateset

The number of Relations
(b) YAGO Dateset

Fig. 1. Data sparsity in YAGO and DBpedia datasets.

- We proposed SM Storage (Sparse Matrix-based Storage) a predicate-based hash index storage, which lifts the storage efficiency.
- We have improve SPARQL join query evaluation over large-scale RDF data by employing sparsity. The RDF data represented by graph, which classified according to the relationships of edges, afterwards the data represented by each relation. Relations are stored in the form of a sparse matrix.
- We transformed the problem of SPARQL query processing to the multiplication operations of multiple sparse matrices, and each triple pattern corresponds to a sparse matrix in the SPARQL query. A sparse matrix is a kind of matrix which consists of large numbers of zero values with very few dispersed non-zero ones. Comparing to general matrices, sparse matrices characterize the sparsity of RDF data well.
- We compared our approach with existing methodologies. It significantly improve BGP query evaluation as compared to others, especially over practical RDF data with 500 million triples.

2   Preliminaries

In this section, we review the work related to our demonstration, which we believe are most related to SMat-J.

2.1   RDF and SPARQL

Let U be a countable set of constants. Let U_s, U_p, and, U_o be three subsets of U. An RDF triple is composed of (*subject, predicate, object*) (for short, (s, p, o)) where *subject* $\in U_s$ denotes an entity or a class of resources; *predicate* $\in U_p$ denotes attributes or relationships between entities or classes; and *object* $\in U_o$ is an entity, a class of resources, or a literal value. An RDF dataset D is a set of RDF triples. An RDF dataset represents a labeled directed graph where predicate is taken as a label, so also called RDF graph. In the sense, we assume that $U_p \cap (U_s \cup U_o) = \emptyset$, predicate are never entities or classes.

Let N_s, N_p, and N_o denote the set of subject, predicates and objects, respectively. Let D be an RDF dataset. Let $N_s(D)$, $N_p(D)$, and $N_o(D)$ denote the set of subjects, predicates, and object occurring in D, respectively. Formally, we define $N_s(D)$, $N_p(D)$, and $N_o(D)$ as follows:

- $N_s(D) := \{u \mid \exists v \in U_p, w \in U_o, \ (u, v, w) \in D\}$;
- $N_p(D) := \{v \mid \exists u \in U_s, w \in U_o, \ (u, v, w) \in D\}$;
- $N_o(D) := \{w \mid \exists u \in U_s, v \in U_p, \ (u, v, w) \in D\}$.

SPARQL is an official recommended RDF query language [6]. SPARQL query language is based on triple patterns, so-called $t_p = (s, p, o)$ and there may be variables in any position, such as:
$?A, : follows, ?B$. And a set of triple patterns forms a basic graph pattern (BGP).

A common SPARQL query contains a group of BGP queries, whose conjunctive fragment allows to express the core form:

$$Select|Project|Join.$$

Given a SPARQL query, we can parse it into query graph where each triple is taken as an edge labelled by its predicate and two edges are connected by their common constant or variable. For example, considering a query as follows: for the above SPARQL query (Fig. 2(b)), the SPARQL query is represented as a query graph (or query pattern) in Fig. 2(b).

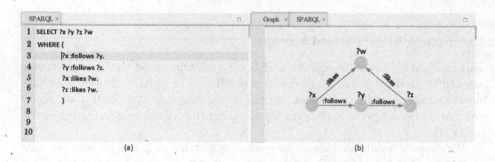

Fig. 2. An example of a SPARQL query.

In addition, each triple pattern can mapped to a sparse matrix. Predicate in each triple is an edge relationship which can be indexed to refer a sparse matrix. A SPARQL query processing problem is about a join operation of multiple triples is transformed into multiplication of sparse matrices. In other words, we transform the SPARQL query problem into a sparse matrix-based multiplication.

3   Framework of SMat-J

In Fig. 3 bigger picture of SMat-J is elaborated, we only covered the partial part of it here in this demonstration. The grey parts of the diagram deals with GPU and will be completed in our future work.

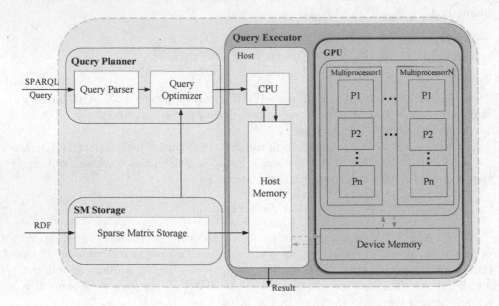

Fig. 3. The framework of SMat-J.

3.1   Sparse Matrix-Based Storage

This module maintains a collection of Sparse Matrix-based (SM-based) tables indexed by predicates and the statistics of predicates from RDF dataset. The SM-based storage provides a compact and efficient method to store RDF data physically and also supports high-performance algorithm of query execution (shown in Fig. 4).

We develop a data dictionary by encoding a string of raw data into positive integers so that we can construct two bijective mappings h_p and h_{so}. The goal of this encoding is to reduce space usage and disk I/O. Secondly, we store RDF Cube (a logical model of SM-based storage) as sets of matrices via a index of predicates. Finally, a series of sparse matrices are generated from the RDF cube indexed by predicates.

3.2   Query Planner

This module has two submodules, Query Parser and Query Optimizer. The query parser parses a SPARQL query as a query graph (shown in Fig. 5) and the latter

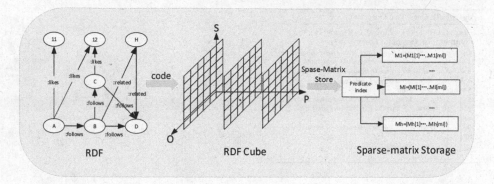

Fig. 4. The framework of SM storage.

generates the optimal query plan on the bases of SM-based storage which owns a sparse matrix (shown in Fig. 6). The cost of is directly proportional to the no of matrices (the intermediate results). The Simple approach is adopted, if no of matrices are greater, the cost of the query will be greater. The query planner passed the optimal plan to query execution module by calculating the query cost. The left side of Fig. 6 is the query plan algorithm, and the right side of Fig. 6 is the generation of query plan graph.

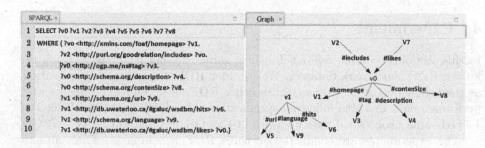

Fig. 5. An example of transforming a query into a graph.

3.3 Query Executor

It is well known that joins of relations and matrix multiplication is essentially the same problem. The entire reason is that both are essentially reachability in two steps in a graph. The SPARQL query is transformed into the multiplications of a series of sparse matrices. Therefore, table-based join operations are converted to sparse matrix-based join (SM-based join) that approximates multiplications for sparse matrices.

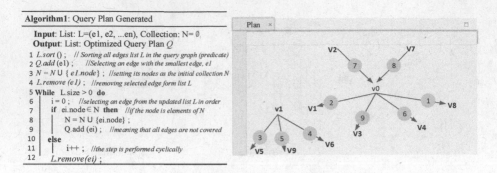

Fig. 6. Generating the query plan.

After obtaining an optimized query, a series of sparse matrices can be obtained based on the predicates of each triple pattern in the query (assuming the predicate is not a variable, which is known). A SPARQL query is a continuous SM-based join operation. For a whole SPARQL query, we need to loop a join operation in a query plan.

Matrix-based multiplication depends on our sparse matrix multiplication algorithm. Due to the length of the article, it is not described here. Experiments show that the sparse matrix proportionality can effectively reduce the intermediate results of the query and increase the query efficiency.

4   Experiments

In this section we evaluate SMat-J against some existing RDF stores using the known RDF benchmark datasets. We consider RDF-3X, gStore for evaluation, as they show better performance than other RDF store. Our experiments were performed on a machine following specifications, Intel(R) Xeon(R) E5-2603 v4 1.7 GHz amd processor, Linux ubuntu 14.04 as OS and main memory is 72 GB.

The experiments are conducted on three RDF benchmarks. The synthetic data sets come from well-established Semantic Web benchmarks: the Waterloo SPARQL Diversity Test Suite (WatDiv [7]) and the real world datasets like open source YAGO and DBpedia. We use five different sizes of synthetic data, which are 100 million, 200 million, 300 million, 400 million and 500 million RDF triples. YAGO is a real RDF dataset which consists of facts extracted from Wikipedia and integrated with the WordNet thesaurus. DBPedia is another real RDF set, which constitutes the most important knowledge base for the Semantic Web community. The data characteristics are summarized in Table 1.

We test the performances of our method in two aspects: the efficiency and the scalability. The WatDiv benchmark defines 20 query templates classified into four categories: linear(L), star(S), snowflake(F) and complex queries(C). We analyze the query efficiency according to the average query time of the four types. In order to avoid the effect of caches on experimental results, we drop the caches before each execution. Query times are averaged over 10 consecutive runs.

Table 1. Benchmark statistics

#Dataset	#Triples	#(S ∩ O)	#P
WatDiv100M	108 997 714	10 250 947	86
WatDiv200M	219 783 842	20 296 483	86
WatDiv300M	329 827 477	30 221 812	86
WatDiv400M	439 433 765	40 040 420	86
WatDiv500M	549 246 141	49 771 433	86
Yago	200 737 655	38 734 252	46
DBpedia	120 978 080	42 966 066	4 282

All results are rounded to 1 decimal place. gStore cannot handle datasets of more than 300 million triples in our environment, so WatDiv400M and WatDiv500M over gStore cannot be included in the results of the experiments.

Table 2. Response time for WatDiv100M (ms)

Wat100	C	F	L	S
SMat-J	3378.1	3784.1	1932.5	2503.3
RDF-3X	11980.3	8405.6	16252.1	3820.6
gStore	15447.0	29204.8	20128.0	10808.7

Table 3. Response time for WatDiv300M (ms)

Wat300	C	F	L	S
SMat-J	18106.2	9185.3	5033.3	5350.3
RDF-3X	39571.1	26231.3	56879.2	16066.5
gStore	111490.8	1047971.5	311390.4	2996687.5

Firstly, we analyze the query efficiency according to average query time for four types oF queries. Due to page limit, we only show partial results (100M, 300M, 500M) on different query engines as shown in Table 2, 3 and 4 respectively.

The results show that SMat-J performs much better than RDF-3X and gStore for all queries. We first considered RDF-3X for comparison. The most obvious acceleration of effect is the L type queries, up to 11 times for SMat-J. For complex type queries, the speedup is generally stable at about 2 times for SMat-J. For other type queries (star and snowFlake), the speedup is generally stable at about 3 times for SMat-J. Next we discuss the comparison with gStore. Due to our environmental constraints, we can only analyze gStore with triple size up to

Table 4. Response time for WatDiv500M (ms)

Wat500	C	F	L	S
SMat-J	35509.1	16071.4	9719.6	9072.4
RDF-3X	66513.7	46906.9	92685.6	27456.7

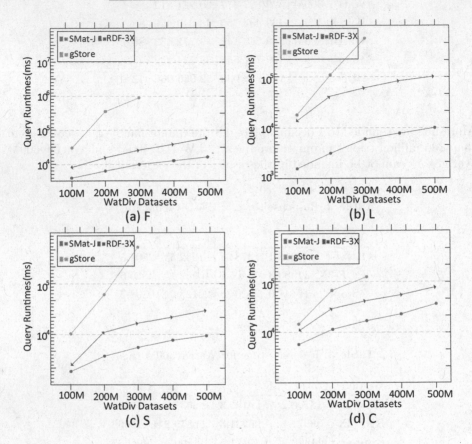

Fig. 7. Average query runtime for WatDiv (ms).

300 million. From the experimental result, we can see that SMat-J has a great improvement for different type queries of data in gStore. In our environment, when the data is a maximum of 300 million, the acceleration ratios of snowflake, linear, star, and complex types are 114.09, 61.87, 55.45, 6.16 for SMat-J.

A comparison graph of the average query time for different types of WatDiv data on different query engines (SMat-J, gStore, RDF-3X) is shown in Fig. 7. As we can see, query results are analyzed on the same data but on different data scales. Figures 7(a), 7(b), 7(c) and 7(d) show the trend of the query time of the F-type, L-type, S-type and C-type of watdiv data in different data scale. With the continuous increase in the scale of data, various types of queries have also increased in time, and they have shown steady and gradual growth. Obviously, it can be seen that SMat-J's query timeline is lower than that of RDF-3X and gStore. Under any type of different data scale, the query time is always at the bottom. Also, as the data size increases, the growth rate of the query time of SMat-J is also the slowest. From the graph, we can also analyze that the query time of gStore increases rapidly for each additional 100M of data, which is very easy to explain the data limit problem of gStore. Because it acquire large system memory for large scale RDF datasets.

Secondly, we test the performances of our SMat-J by using the real datasets, YAGO and DBpedia. Figure 8 describes a comparison among the query runtime of real datasets YAGO and DBpedia on different query engines (SMat-J, gStore, RDF-3X). We discuss query execution performance on both datasets separately. First we analyze the comparison on YAGO data as shown in Fig. 8(a). Due to YAGO does not provide benchmark queries, we design seven queries for YAGO real datasets [8]. And it covers the four query types mentioned above. As can be seen, the minimum query time is SMat-J. This fully shows that our SMat-J has a very good scalability.

As can be seen, compared with RDF-3X, the most obvious acceleration effect for SMat-J is the Q3 (F-type), up to 33.6 times. Overall, Q2 and Q3 are the most efficient comparisons of RDF-3X. And compared with gStore, the speedup for SMat-J is the Q6 (S-type), up to 46.4 times. Overall, Q6 and Q7 are the most efficient comparisons of gStore.

Then we analyze the comparison on DBpedia data as shown in Fig. 8. For DBpedia dataset, we design four queries. Analysis of query efficiency on DBpedia data is similar to YAGO. As can be seen from the figure, the time of SMat-J is smaller than gStore and RDF-3X.

Overall, for query efficiency analysis of real data, SMat-J is significantly more efficient than RDF-3X and gStore.

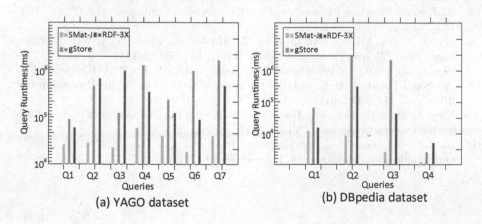

(a) YAGO dataset   (b) DBpedia dataset

Fig. 8. Average query runtime for YAGO and DBpedia (ms).

5   Conclusion

By visualizing above experimental results and discussion, it is concluded that SMat-J has outclassed the RDF3X and gStore, especially, for large scale RDF datasets. In the future we work on a remaining part of SMat-J framework. We will execute SMat-J on GPU and in distributed environment with an engagement of parallelism.

References

1. Manola, F., Miller, E.: RDF Primer. W3C Recommendation, 10 February 2004
2. Abdelaziz, I., Harbi, R., Khayyat, Z.: A survey and experimental comparison of distributed SPARQL engines for very large RDF data. PVLDB **10**(13), 2049–2060 (2017)
3. DBpedia. http://dbpedia.org/
4. YAGO2. http://yago-knowledge.org/
5. Atre, M., Chaoji, V., Zaki, M.J., Hendler, J.A.: Matrix "Bit" loaded: a scalable lightweight join query processor for RDF data. In: Proceedings of WWW, pp. 41–50 (2010)
6. Perez, J., Arenas, M., Gutierrez, C.: Semantics and complexity of SPARQL. ACM Trans. Database Syst. **34**(3), 16 (2009)
7. Aluç, G., Hartig, O., Özsu, M.T., Daudjee, K.: Diversified stress testing of RDF data management systems. In: Proceedings of ISWC, pp. 197–212 (2014). http://dsg.uwaterloo.ca/watdiv/
8. https://raw.githubusercontent.com/zhangmingyue93/zzz/master/index.md
9. Neumann, T., Weikum, G.: The RDF-3X engine for scalable management of RDF data. VLDB J. **19**(1), 91–113 (2010)
10. Yuan, P., Liu, P., Wu, B., Jin, H., Zhang, W., Liu, L.: TripleBit: a fast and compact system for large scale RDF data. PVLDB **6**(7), 517–528 (2013)
11. Brisaboa, N.R., Cerdeira-Pena, A., de Bernardo, G., Fariña, A.: Revisiting compact RDF stores based on k2-trees. In: Data Compression Conference (DCC), pp. 123–132. IEEE (2020)
12. Zou, L., Ozsu, M.T., Chen, L., Shen, X., Huang, R., Zhao, D.: gStore: a graph-based SPARQL query engine. VLDB J. **23**(4), 565–590 (2014)
13. Wylot, M., Pont, J., Wisniewski, M., Cudre-Mauroux, P.: dipLODocus[RDF] - short and long-tail RDF analytics for massive webs of data. In: Proceedings of ISWC, pp. 778–793 (2011)
14. Kim, J., Shin, H., Han, W., Hong, S., Chafifi, H.: Taming subgraph isomorphism for RDF query processing. PVLDB **8**(11), 1238–1249 (2015)
15. Ingalalli, V., Ienco, D., Poncelet, P., Villata, S.: Querying RDF data using a multigraph-based approach. In: Proceedings of EDBT, pp. 245–256 (2016)

A Distributed Engine for Multi-query Processing Based on Predicates with Spark

Bin Zhang[1], Ximin Sun[1(✉)], Liwei Bi[2], Changhao Zhao[2], Xin Chen[2], Xin Li[2], and Lei Sun[3]

[1] State Grid Financial Technology Group,
State Grid Electronic Commerce Co., Ltd., Beijing, China
{zhangbin,sunximin}@sgec.sgcc.com.cn
[2] State Grid Ecommerce Technology Co., Ltd., Beijing, China
{biliwei,zhaochanghao,chenxin1,lixin3}@sgec.sgcc.com.cn
[3] College of Intelligence and Computing, Tianjin University, Peiyang Park Campus,
Tianjin, China
sunlei_2020@tju.edu.cn

Abstract. The core of Multi-query Optimization is to find the maximum common substructure of query graphs. However, this problem is equivalent to the weighted set cover problem, which is NP-complete. In this paper, we propose an distributed RDF engine for multi-query processing with Spark. The system processes SPARQL queries with translating them into Spark SQL. We utilize the predicate information as the feature of the query and cluster the multiple queries which share more common features into groups.

We conduct experiments with synthetic datasets, compared with the result without MQO processing, we could show the effectiveness of our approach.

Keywords: Multi-query · RDF · Spark

1 Introduction

In recent years, with different kinds of information increasing rapidly, the traditional relational data model can not satisfy the producing and entertainments requirements of people gradually. Therefore, W3C recommend the Resource Description Framework (RDF) as the new data model to describe arbitrary resource on the Internet. RDF has become the real standard and been adopted widely by the semantic Web communities and commercial organizations. DBpedia, the important application for Linked Open Data [1], consists of over 1 billion triples. The LOD dataset currently contains 1301 datasets with 16283 links (as of May 2020).

With the rapid development of the semantic data, more and more research begin to focus on the indexing, store RDF data and processing SPARQL queries.

© Springer Nature Singapore Pte Ltd. 2021
Y. Gao et al. (Eds.): APWeb-WAIM 2021 Workshops, CCIS 1505, pp. 27–36, 2021.
https://doi.org/10.1007/978-981-16-8143-1_3

D-SPARQ [2] use MapReduce framework and MongoDB database and will reorder triples with the selectivity of triples. MR-RDF [3] proposes a RDF data partitioning approach, which cluster would entities based on the schema. TriAD [4] is a distributed shared-nothing RDF engine, which present a new RDF summary graph for join-ahead data pruning and utilizing the asynchronous message passing. Partout [5] attaches more importance to the query log and partitioning, which would fragment data based on the simple predicate of query load. gStore-D [6] partitions RDF data into fragments of which vertices are disjoint, evaluate SPARQL queries with partial evaluation and assembly strategy. S2RDF [7] propose a new RDF data partitioning schema called ExtVP, which could reduce the size of the input queries obviously. Meanwhile, it implements a compiler which could translate SPARQL queries into SQLs with statistics supplied by ExtVP. Stylus [8] defines a strong style key-value storage and present an indexing structure xUDT and star shape query unit Twig for further query processing. AdPart [9] utilizes light hash partitioning to distribute triples and dynamically redistributes and replicates the instances of the most frequent ones among workers through monitoring the data access patterns. StarMR [10] proposes an easy scalable algorithm based on star shape decomposition, which would filter the ineffective input data of MapReduce iterations with attributes of RDF graph. DP2RPQ [11] presents a distributed Pregel approach to evaluate RPQs of RDF graph, which designs three optimized strategies, vertex-computation optimization, message-communication reduction, and counting-paths alleviation.

The contribution of the paper can be summarized as follows: (1) a distributed Engine for Multi-query processing built on top of Hadoop-Spark; (2) we conduct a comparative experiment between prototype and our system over WatDiv Dataset via synthetic queries. The experimental results show the effectiveness of our approach.

The rest of the paper is organized as follows. In Sect. 2, we introduce the related work and background about the existing RDF systems. In Sect. 3, we provide an overview of our system, which briefly introduce the system architecture. And we provide a detailed description of the system and working process in Sect. 4. Section 5 mainly introduce the experimental setup, configuration and results. Eventually, we present the conclusions and future work in Sect. 6.

2   Related Work

2.1   RDF

A RDF dataset consists of a collection of triples, which are represented as (subject, predicate, object). In detail, subject means a globally unique resource, object includes a globally unique resource or a literal [12]. And predicate represents the relationship between subject and object. In this way, a triple can describe a fact in real world. Here follow the examples of the RDF triple.

James_Cameron < directs > Titanic
James_Cameron < networth > 1.79E9
Titanic < rdf:type > Movie
Titanic < budget > 2.0E8
Leonardo_DiCaprio < rdf:type > Actor
Leonardo_DiCaprio < acts_in > Titanic
Kate_Winslet < rdf:type > Actor
Kate_Winslet < acts_in > Titanic

We show an example of RDF dataset in Fig. 1, where rectangular nodes mean liteal value, ellipse nodes represent resources, direct edges denotes the properties between entities.

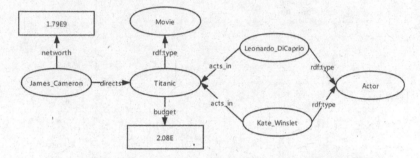

Fig. 1. An RDF graph example of movie knowledge graph.

2.2 Multi-query Processing

Multi-query optimization (MQO) is a classical problem for RDF and SPARQL queries. First of all, there exist some researches on MQO problems on traditional relational databases and semi-structure data [13]. Hong et al. presents a Massively Multi-Query Join Processing techniques, of which the key is to explore the join conditions into tree pattern and value comparison components, utilizing the sharing of representations of inputs of multiple joins and the sharing of join computation [14]. For multi-query optimization in the context of RDF and SPARQL, Kementsietsidis A et al. proposes an two-stage heuristics to optimize the multi-queries [15]. Liu et al. [16] proposes the HadoopSPARQL system, which could detect common subqueries and use multi-way join operator based on MapReduce framework. Le et al. [17] also adopt the heuristic algorithm to partition queries into groups for further optimization. And the optimization method will include efficient algorithm to detect the common sub-structures of multiple queries and cost model for determining execution plans.

3   System Overview

An overview of the system is shown in the Fig. 2. The System is implemented with Hadoop-Spark framework, and it follows the classical master-slave architecture. The master node holds the Data Loader, which would store all data with the parquet format and produce the predicate statistics based on predicate information. Besides, the master node also holds the Query Translator, which would translate the SPARQL into Spark SQL with a compiler. The slave nodes will engage in querying phrase, they hold the query parser, query feature generator and query executor for multiple queries processing.

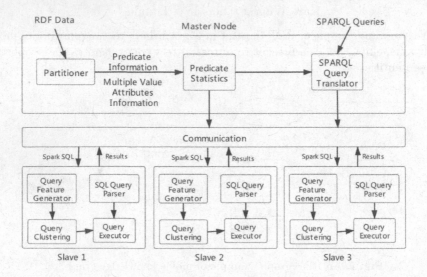

Fig. 2. System architecture.

4 Proposed Approach

In this section, we will describe the complete query processing in details. The system is implemented with distributed Hadoop-Spark SPARQL query processor, which use the interface of Spark to finish querying with translating SPARQL queries into SQL. The detail of our algorithm is shown in Algorithm 1.

4.1 Predicate Statistics

Property table is a traditional method to store triples, however, this method encounter the drawbacks of failure of storing multi-valued properties. Modified Property Table [18] is proposed to fix the problem up, it store multi-valued properties in a single cell with a nested data structure. Based on this structure, an extension named PTP (property table partitioning) [19] is introduced to reduce the number of tuples and I/O costs for queries, which could better process the complex queries like star patterns and so on.

In this process, the system will produce a statistics file, which records the actual size of each PTP table and name of each multi-valued attribute. Furthermore, this statistics file will play an important role in the query generation.

Algorithm 1. Multi-query Optimization

Input: A batch of queries Q={Q_1,Q_2,. . .,Q_n}
Output: Query results
1: Load data and generate statistics file S.
2: Compute clustering cost for different number of clusters
3: Determine the best value for number of clusters K
4: %K-Means clustering
5: C=QueryClustering(Q,k,S)
6: %Query Translator
7: **for** each C_i in C **do**
8:     Translate C_i into sql C_i'
9: %Query Executor
10: **for** each C_i' in C' **do**
11:     Execute C_i' and get results
12: **return** Query results

4.2   Query Translator

This system process SPARQL queries with translating them into SQLs. Generally speaking, the Basic Graph Pattern (BGP) is the most common type for the SPARQL queries. The triple patterns which have same subject are grouped together. Therefore, the BGP are decomposed into several subqueries which have same subjects. During the process, each subquery could be translated into corresponding SQL.

Afterwards, with all subqueries translating to corresponding SQLs, execution order should be taken into consideration to reduce the execution costs. According to the previous statistics file generated in data loading phrase, the actual size of each table of each predicate can be obtained. The subquery with more constrained value and smaller size will have higher priority and be executed earlier. Through join operators, the subqueries will be aggregated together to acquire the final results in the end.

4.3   Query Feature Generator

In order to model the SPARQL queries and cluster them into groups, we need to extract the feature of each SPARQL query. By analysing the structure of the SPARQL queries, the predicate part would be more appropriate to represent the queries.

At first, we should construct a template vector according to the predicate information of the statistics file generated in the data loading phrase. Each position of the vector corresponds to a predicate in the dataset, which are initialized as 0. Then, predicates of each query are extracted to represent each query through set the corresponding position of the vector to 1. In this way, each vector could represent the corresponding query. Besides, this way matches the data storage method of the system based on the PTP schema, which would facilitate the query execution step.

Algorithm 2. QueryClustering

Input: A batch of queries Q={Q_1,Q_2,. . .,Q_n}
The number of clusters k
The statistics of data S
Output: k clusters of queries C={C_1,C_2,. . .,C_k}
1: %Constructing feature vectors for each query Q_i
2: **for** each predicate p_i in Q_i **do**
3:     Set the position of p_i of vector q_i for Q_i to 1
4: %K-Means clustering
5: Choose k centroids randomly from q_i
6: **for** each iteration **do**
7:     **for** each q_i **do**
8:         Compute the distance between q_i and k centroids
9:         Assign each q_i to its closest cluster
10:     **for** each cluster C_i **do**
11:         Compute the average of C_i and update it as new centroid
12: **return** C

4.4   Multiple Queries Clustering

In this step, with query modeling into vectors, we need to cluster queries into groups for future query execution. The detail of our algorithm is shown in Algorithm 2.

K-Means is a classical clustering algorithm with easy implementation and good results. The algorithm has four steps, the first step is to choose k centroids randomly from all samples. The second step is to compute the distance between each sample and each centroid and assign each sample to its closest centroid. The third step is to compute the average value of each cluster and update as new centroids. The final step is to repeat the second and third step until the iterations finish or results converge.

However, there exists a drawback for K-Means algorithm. The number of cluster k should be set manually, which is not easy to figure out a proper digit. Actually, we implement the K-Means algorithm with the help of the class KMeansModel of Spark MLlib, and it provides a function computeCost which could compute the square of distance between all the samples and their closest centroids to evaluate the performance of clustering. In fact, choosing the result with minimal cost is not always a good option, realistic situation should be taken into consideration.

With queries vectors, we use K-Means algorithm cluster them into several groups share more common predicates. Then we translate queries of each group into SQLs with compiler.

4.5   Query Executor

In this section, the system will utilize the relational interface of the Spark to execute SQLs. For seriousness, each SQL will be execute four times and compute means as the result in order to avoid coincidence.

The queries belonging to same group share more common predicates, when executing queries in one batch, the triples of predicates corresponding to this cluster centroid will be preteched into the memory. Therefore, the batch queries will not need to repeat loading the same triples, which could save the time and costs.

5   Experiment

In this section, we conduct a comparative experiment with the system without multi-query processing. And the experimental setup and discussion of the results are demonstrated.

5.1   Benchmark Queries

In order to evaluate the performance of the management solution to the multi-query processing, the synthetic dataset is utilized and shown in Table 1.

Table 1. Experimental setup and dataset scale.

Dataset	Scale factor	Number of triples	Number of queries
WatDiv	1	10k	200

The synthetic dataset WatDiv is a benchmark and artificial dataset of semantic dataset, introduced by DATA SYSTEM GROUP of the University of Waterloo in 2014. Moreover, it has both data generator and query generator, which means it could generate different kinds and sizes of SPARQL queries according to the input. There are four types of queries, which are Linear (L), Star (S), Snowflake (F), Complex Query (C), just as the name suggests. Every query group has different members, as shown in the following:

- Linear shape (L): L1, L2, L3, L4, L5.
- Star shape (S): S1, S2, S3, S4, S5, S6, S7.
- Snowflake shape (F): F1, F2, F3, F4, F5.
- Complex Query (C): C1, C2, C3.

5.2   Experimental Configuration

We construct a distributed environment with three nodes (1 master and 2 workers) on the virtual machines. The master node has 8 GB of memory and 50 GB of hard disk space, and each worker node has 2 GB of memory and 50 GB of hard disk space. More importantly, Parquet filter pushdown of each node is enabled.

5.3   Experimental Results

We present a comparative experiment of our system with the others which do not support multi-query processing to show the effectiveness of our approach. In data loading process, there are two phrases. At first, we need to create the Property Table. On the basis of this step, we need to create the PTP (property table partitioning) tables. And the time are not included in the query execution time. From the Table 2, we notice that the system with multi-query processing accelerate the query process. When the system execute the queries belonging to one group, it will prefetch the information of the center predicate of this group. Therefore, these queries only need to load their own relevant data according to their predicates into memory. In Fig. 3, results of executing 200 queries are shown. By this way, the system could save time for loading the common part of these queries in one group, meanwhile, the memory will be saved for avoiding replicate data.

Table 2. WatDiv query runtimes.

System	Query runtimes
Prototype	4 min
Multi-query processing	2.25 min

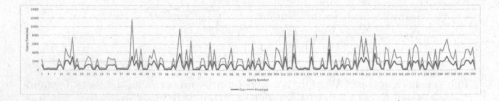

Fig. 3. Results of executing 200 queries.

6   Conclusion and Future Work

In this paper, we propose an approach to the multi-query processing and implement a distributed RDF storage and query system based on Hadoop and Spark. This method fully utilize the predicate information of the queries to explore the common part of the multiple queries. And we model the SPARQL queries with their predicate information and cluster them into groups based on the structure of queries. We conduct a comparative experiment between the prototype and our system. The experimental results demonstrate that our approach could accelerate the multiple queries.

For future work, we consider more improvements of the system to enhance the usability and query performance. Meanwhile, we are investing useful methods of reducing costs of loading data and executing queries. Considering the massive matrix computation the system involved with, GPU acceleration will be a promising aspect to improve the performance.

Acknowledgements. The authors appreciate the equipment and environment supported by the STATE GRID ELECTRONIC COMMERCE CO., LTD./STATE GRID FINANCIAL TECHNOLOGY GROUP and STATE GRID ECOMMERCE TECHNOLOGY CO., LTD.

References

1. Bizer, C., Heath, T., Berners-Lee, T.: Linked data - the story so far. Int. J. Semantic Web Inf. Syst. **5**(3), 1–22 (2009)
2. Mutharaju, R., Sakr, S., Sala, A., Hitzler, P.: D-SPARQ: distributed, scalable and efficient RDF query engine. In: International Semantic Web Conference (Posters & Demos), pp. 261–264 (2013:)
3. Fang, D., Bian, H., Chen, Y.: Efficient SPARQL Query Evaluation in a Database Cluster, pp. 165–172. BigData Congress, Xiaoyong Du (2013)
4. Gurajada, S., Seufert, S., Miliaraki, I., Theobald, M.: TriAD: a distributed shared-nothing RDF engine based on asynchronous message passing. In: SIGMOD Conference, pp. 289–300 (2014)
5. Galárraga, L., Hose, K., Schenkel, R.: Partout: A Distributed Engine for Efficient RDF Processing. WWW (Companion Volume), pp. 267–268 (2014)
6. Peng, P., Zou, L., Tamer Özsu, M., Chen, L., Zhao, D.: Processing SPARQL queries over distributed RDF graphs. VLDB J. **25**(2), 243–268 (2016)
7. Schätzle, A., Przyjaciel-Zablocki, M., Skilevic, S., Lausen, G.: S2RDF: RDF querying with SPARQL on spark. Proc. VLDB Endow. **9**(10), 804–815 (2016)
8. He, L., et al.: Stylus: a strongly-typed store for serving massive RDF data. Proc. VLDB Endow. **11**(2), 203–216 (2017)
9. Al-Harbi, R., Abdelaziz, I., Kalnis, P., Mamoulis, N., Ebrahim, Y., Sahli, M.: Accelerating SPARQL queries by exploiting hash-based locality and adaptive partitioning. VLDB J. **25**(3), 355–380 (2016)
10. Wang, X., et al.: Efficient subgraph matching on large RDF graphs using mapReduce. Data Sci. Eng. **4**(1), 24–43 (2019)
11. Wang, X., Wang, S., Xin, Y., Yang, Y., Li, J., Wang, X.: Distributed Pregel-based provenance-aware regular path query processing on RDF knowledge graphs. World Wide Web **23**(3), 1465–1496 (2020)
12. RDF 1.1. https://www.w3.org/TR/rdf11-concepts
13. Guo, X., Gao, H., Zou, Z.: Leon: a distributed RDF engine for multi-query processing. In: Li, G., Yang, J., Gama, J., Natwichai, J., Tong, Y. (eds.) DASFAA 2019. LNCS, vol. 11446, pp. 742–759. Springer, Cham (2019). https://doi.org/10.1007/978-3-030-18576-3_44
14. Hong, M., Demers, A.J., Gehrke, J., Koch, C., Riedewald, M., White, W.M.: Massively multi-query join processing in publish/subscribe systems. In: SIGMOD Conference, pp. 761–772 (2007)
15. Kementsietsidis, A., Neven, F., Van de Craen, D., Vansummeren, S.: Scalable multi-query optimization for exploratory queries over federated scientific databases. Proc. VLDB Endow. **1**(1), 16–27 (2008)
16. Liu, C., Qu, J., Qi, G., Wang, H., Yu, Y.: HadoopSPARQL: a hadoop-based engine for multiple SPARQL query answering. In: ESWC (Satellite Events), pp. 474–479 (2012)

17. Le, W., Kementsietsidis, A., Duan, S., Li, F.: Scalable multi-query optimization for SPARQL. In: ICDE 2012, pp. 666–677 (2012)
18. Hassan, S., Bansal, SK.: Data partitioning scheme for efficient distributed RDF querying using apache spark. In: IEEE 13th International Conference on Semantic Computing (ICSC), pp. 24–31 (2019)
19. Hassan, M., Bansal, S.K.: S3QLRDF: property table partitioning scheme for distributed SPARQL querying of large-scale RDF data. In: 2020 IEEE International Conference on Smart Data Services (SMDS). IEEE (2020)
20. Meimaris, M., Papastefanatos, G., Mamoulis, N., Anagnostopoulos, I.: Extended characteristic sets: graph indexing for SPARQL query optimization. In: ICDE 2017, pp. 497–508 (2017)

Product Clustering Analysis Based on the Retail Product Knowledge Graph

Yang Ye[1,2] and Qingpeng Zhang[1,2(✉)]

[1] School of Data Science, City University of Hong Kong, Hong Kong SAR, China
yang.ye@my.cityu.edu.hk, qingpeng.zhang@cityu.edu.hk
[2] The Laboratory for AI-Powered Financial Technologies, Hong Kong SAR, China

Abstract. Product clustering analysis is essential in designing retail marketing strategies. It is a common practice that retailers use to effectively manage their product inventory, marketing promotions, etc. The most intuitive way of clustering products is by their explicit attributes, such as brand, size, and flavor. However, these approaches do not integrate the customer-product interactions, thus ignore the implicit product attributes. In this work, we construct a retail product knowledge graph based on Amazon product metadata. Leveraging a state-of-the-art network embedding method, RotatE, our main objective is to unveil hidden interactions of products by including implicit product attributes. These hidden interactions bring insights to downstream operations such as demand forecasting, production planning, assortment optimization, etc.

Keywords: Clustering · Retail product knowledge graph

1   Introduction

Applying data mining and machine learning techniques has been a growing trend in the retail industry recently. Leveraging high-level data and information, these techniques enable retailers to better understand customers' behaviors, design powerful inventory forecasting strategies, optimize retail operations, etc. [2,6,7]. Clustering algorithms are unsupervised machine learning algorithms, which are utilized to automatically cluster unlabeled data points into multiple groups based on their similarities. Product clustering is a common practice in the retail industry that retailers use to effectively manage their product inventory, marketing promotions, etc. Most commonly, explicit product attributes, such as brand, size, and flavor, are used to cluster products [4]. More advanced approaches leverage text data (e.g., description of products) and image data of products for clustering [1]. These methods do not integrate information of customer-product interactions, such as the co-occurrence of products in the same order, thus ignore

The work described in this paper was partially supported by Laboratory for AI-Powered Financial Technologies.

© Springer Nature Singapore Pte Ltd. 2021
Y. Gao et al. (Eds.): APWeb-WAIM 2021 Workshops, CCIS 1505, pp. 37–40, 2021.
https://doi.org/10.1007/978-981-16-8143-1_4

some implicit product attributes. Here, we first learn the implicit attributes of products by adopting the knowledge graph embedding approach; then we use the Principal Component Analysis (PCA) method to transform data from high-dimensional feature space to lower-dimensional space in which to perform clustering analysis. We find that product clustering results based on implicit product attributes bring up insights about the hidden relations between products.

2   Results

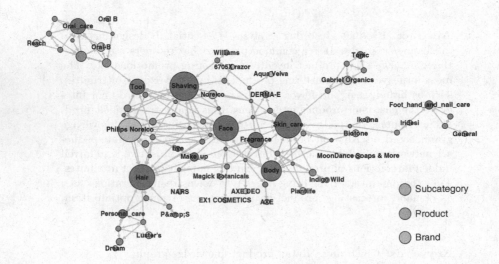

Fig. 1. Subgraph from the complete product knowledge graph.

The dataset used in our experiments is constructed from existing Amazon product metadata [3] that includes descriptions, price, sales rank, brand info, and co-purchasing links for 15.5 million products. We select 2780 products from "All Beauty" category and construct the product knowledge graph based on the following attributes in the metadata: "asin", "brand", "title", "description", "feature", "similar_item", "also_buy", and "also_view". The product knowledge graph stores entities (subcategories, products, and brands) and inter-entity relations (has_the_brand, in_the_subcategory, similar_item, also_buy, and also_view) in a triple format, (h, t, r), where h, t, and r denote the head entity, the tail entity, and the relationship between h and t, respectively. Note that, "title", "description", and "feature" are used to infer the subcategories of products, which are not provided in the metadata. Figure 1 shows a subgraph of the complete product knowledge graph. We adopt RotatE [5], an approach for knowledge graph embedding, to learn the implicit attributes of products. Figure 2 illustrates the visualization of embeddings for products in different subcategories. We can find that, even without a clustering loss, RotatE automatically embeds products in the same subcategory into the same cluster; products in more related subcategories

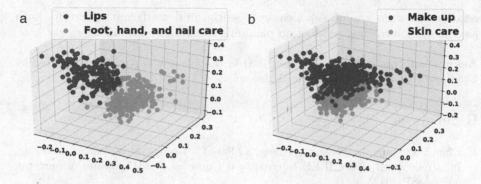

Fig. 2. Visualization of embedding for products in different subcategories.

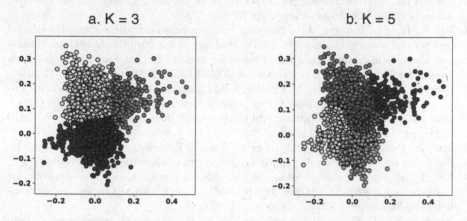

Fig. 3. K-means clustering results.

are closer in the vector space. We then run the K-means algorithm to get product clustering results. Figure 3 shows the clustering results for $K = 3$ and $K = 5$. Each cluster contains not only products in a few dominant subcategories but also those in several related subcategories. By taking customer-product interactions into consideration, these clustering results can help retailers to discover hidden relations between products and to improve their existing assortment strategies.

3 Conclusion

In this work, we perform a product clustering analysis based on the implicit attributes of products. We use a knowledge graph embedding approach, RotatE, to learn the implicit attributes of products. The product knowledge graph is constructed using product metadata from Amazon. The metadata contains not only information of explicit product attributes (such as brand and subcategory) but also the information of customer-product interactions. We find that product clustering results based on the implicit product attributes unveil the hidden relations between products, which are essential in downstream operations such as demand forecasting, production planning, assortment optimization, etc.

Acknowledgements. The work described in this paper was partially supported by Laboratory for AI-Powered Financial Technologies.

References

1. Chang, J., Wang, L., Meng, G., Xiang, S., Pan, C.: Deep adaptive image clustering. In: Proceedings of the IEEE International Conference on Computer Vision, pp. 5879–5887 (2017)
2. Lucic, A., Haned, H., de Rijke, M.: Why does my model fail? Contrastive local explanations for retail forecasting. In: Proceedings of the 2020 Conference on Fairness, Accountability, and Transparency, pp. 90–98 (2020)
3. Ni, J., Li, J., McAuley, J.: Justifying recommendations using distantly-labeled reviews and fine-grained aspects. In: Proceedings of the 2019 Conference on Empirical Methods in Natural Language Processing and the 9th International Joint Conference on Natural Language Processing (EMNLP-IJCNLP), pp. 188–197 (2019)
4. Riaz, M., Arooj, A., Hassan, M.T., Kim, J.B.: Clustering based association rule mining on online stores for optimized cross product recommendation. In: The 2014 International Conference on Control, Automation and Information Sciences (ICCAIS 2014), pp. 176–181. IEEE (2014)
5. Sun, Z., Deng, Z.H., Nie, J.Y., Tang, J.: Rotate: Knowledge graph embedding by relational rotation in complex space. arXiv preprint arXiv:1902.10197 (2019)
6. Xia, L., et al.: Knowledge-enhanced hierarchical graph transformer network for multi-behavior recommendation. In: Proceedings of the AAAI Conference on Artificial Intelligence, vol. 35, pp. 4486–4493 (2021)
7. Yang, S., et al.: Financial risk analysis for SMEs with graph-based supply chain mining. In: IJCAI, pp. 4661–4667 (2020)

The Third International Workshop on Semi-structured Big Data Management and Applications

Survival Effect of Internet Macroscopic Topology Evolution

He Tian[1,2], Kaihong Guo[1(✉)], Zheng Wu[1], and Mingxi Cui[1]

[1] Liaoning University, Shenyang 110036, China
[2] Liaoning Institute of Science and Technology, Benxi 117004, China

Abstract. From the point of biological evolution, this paper studies the survival effect of IPv4 and IPv6 in the evolution, which provides a richer research method and a larger research space for the future Internet. The IPv4 and IPv6 whole network data from 2011 to 2015 authorized by CAIDA is adopted in this paper. The self-organizing metabolism of the Internet is measured by the standard network structure entropy, and the self-replicating behaviors of the Internet are represented by the changes of the degree distribution rates in K-core network. Statistically analyze the survival mutation by the birth rate and the death rate, and predict network transmission mutation by the changes of the average path length. It shows that both of IPv4 and IPv6 Internet topology have vital signs, they emerge an order in the evolution process, in addition, IPv6 has greater order and metabolic capacity. As the stripping goes deeper and deeper, the self-replicating capacity of IPv4 is weakened from the outer layer to the inner layer, but is not affected in IPv6. The changes of the dead rate and the average path length in IPv6 are more active than those in IPv4, and the propagation rate in IPv6 is faster than that in IPv4. The self-replication and information propagation of the network topology in IPv6 are prone to mutation. All of the conclusions provide theoretical support for predicting and guiding evolution direction of the Internet.

Keywords: Vital signs · Internet evolution · Topology

1 Introduction

According to the Darwinian evolution theory, in the nature, biological heredity, variation and natural selection can lead to adaptive changes; organisms can metabolize spontaneously in order to survive and reproduce to make one's own physiology or structure transformation with the change of external environment [1]. In essence, a lift system, which maintains the survival and stability

Supported by the National Natural Science Foundation of China under Grant 71771110 and the Planning Research Foundation of Social Science of the Ministry of Education of China under Grant 16YJA630014 and the Basic Scientific Research Project of Education Department of Liaoning Province (Surface Project) under Grant LJKZ1065.

Y. Gao et al. (Eds.): APWeb-WAIM 2021 Workshops, CCIS 1505, pp. 43–53, 2021.
https://doi.org/10.1007/978-981-16-8143-1_5

of the system through metabolic activities, is only the carrier of life information movement [2]. For the Internet in the human world, other social systems
around it constitute the ecological environment for its survival [3]. Internet evolution makes the interaction and influence among its internal organizations and
other external social environments. Internet constantly destroys the balance of
its own systems while steadily chooses to generate new typologies. Thus it can
be seen that Internet is an open system far away from equilibrium [4]. In order
to evolve stably for a long time, it must be metabolized substances, energy and
information, whereas, metabolic activities will inevitably lead to the reorgani-
zation (self-replication) and variation (mutation) of its own topology to adapt
to the changing environment, which is the trajectory of life information on the
Internet. Regard the Internet as an open life system in view of Langton's "artificial life" idea is a bold attempt in the field of complex network research [5].
Although some literatures in recent years have discussed the vital signs presented
in the Internet evolution process and attempted to extract the life logic from its
macro topology [6], it has not received widespread attention and recognition.
Kim, et al. [7] analyzed the fractal characteristics and the formation reasons of
the network, explained the self-replicating behavior using fractal self-similarity
in the complex network, however, Schneider, et al. [8] considered that the conclusion was not reasonable. Liu, et al. [9] analyzed the dissipative structure of
the Internet evolution order that can integrate organic life, revealed the fluctuation phenomenon in the dissipative structure of the Internet and balanced
those small fluctuations through metabolic activities to maintain the system
stability. Tian, et al. [10] pointed that network traveling time is presented on
multi-peak and heavy tail distribution during the evolution of Internet topology.
Previous studies only reflected the vital signs of the Internet at a certain stage
of the phenomenon. They did not analyze synthetically the whole life process of
the Internet without systematic evaluation. Based on the above research background, this paper introduces the idea of biological evolution into the Internet
for research as follows:

(1) Contrast and analyze the metabolic activities, self- replicating behavior and
    mutation effect in the evolution of IPv4 and IPv6 Internet topology. Observe
    the evolution of standard network structure entropy to analyze the order and
    metabolic characteristics of Internet topology;
(2) Analyze 1–25 cores, then contrast and analyze the distribution similarity
    of degree distribution rate of IPv4 and IPv6 in different years, explore the
    self-replicating behavior of Internet evolution;
(3) Calculate respectively the birth rate and death rate of nodes in IPv4 and
    IPv6, analyze their changing trends to research the mutation caused by
    self-organization metabolism and replication in Internet topology. Analyze
    the mutation effect at that point of the network operation and information
    spreading using network average path length.

Synthesize the conclusions and evaluations obtained by the performance characteristics of the survival effect in the three stages.

2   Theoretical Basis and Data

Material systems usually have some physical properties to adapt to self-organizations and metabolic activities, transiting from high-energy reactants to low-energy products and self-replication. Unlike the transmission and transfer of information in noisy channels, each genetic replication should preserve the meaning of information as far as possible, meanwhile, errors in self-organization cycles are also necessary because of some changing factors, which cause part of the information not to be duplicated correctly. According to the view of modern biology, only systems with metabolism, self-replication and mutation behavior have vital signs. Therefore, the survival effect of life systems in the process of evolution is defined as follows:

Definition 1. *Survival effect [11]: It is a phenomenon that occurred when a system was making self-adaption regulation constantly, of which the evolution process produced information metabolism, self-replication and mutation.*
Metabolism: It is the operation and information transmission process of the whole system, which reflects the fusion ability of topology.
Self-replication: It is a self-organizing behavior in order to keep general information from being lost when a system is evolving.
Mutation: It is the error that caused by noise or other external factors when a system is in the self-replication, and is the motive power for a system development.

Definition 2. *Network structure entropy [12]: It describes the order extent of the topology states of complex network macroscopically. It is also a measure that the chaotic extent of system components. Weigh the possibility of internal dynamic development under the micro state of the system. Its calculation formula is as follows:*

$$H = -\sum_{k=1}^{N} p(k) \ln p(k) \tag{1}$$

Among them, $p(k)$ is the degree distribution rate of nodes in the network. k is node degree. N is the total number of nodes in the network topology.

Definition 3. *Standard structure entropy [13]: It refers to the normalization of network structure entropy, which the calculation formula is changed as follows:*

$$H_s = \frac{H - H_{\min}}{H_{\max} - H_{\min}} \tag{2}$$

Standard structure entropy has nothing to do with number of selected data samples, which is more conducive to further explore the metabolic behaviors and evolution trends of network macro topology.

Definition 4. *K-core [14]: It refers to the surplus subgraphs after repeatedly removing the nodes whose degree is less than or equal to k and their conjoint edges in the network topology. It can be used to describe the hierarchical structure of network.*

Definition 5. *Average path length [15]: It refers to the mean of the shortest path length between any two nodes in the network, which expressed by the formula as follows:*

$$L = \frac{1}{N(N-1)} \sum_{i,j} d_{ij} \qquad (3)$$

Where d_{ij} is the shortest path length from node i to node j. Average shortest path is an important indicator of network communication ability and transmission performance. The smaller the average path length is, the faster the information transmission speed is.

In this paper, the data comes from CAIDA detection project. Select the macro topology data of IPv4 and IPv6 Internet from 2011 to 2015 by taking the month as the detection unit, 60 months in total, to explore the survival effect in the evolution of Internet topology under these two protocols. Barabsi, et al. [16] have proved that using random sampling data instead of all the data is reliable for research results. We can research the essence of the evolution of Internet entropy by measuring a small number of probe sources.

3   Metabolic Analysis of Internet Topology

The chaotic extent of the Interwork structure is directly related to the exchanges, transmission and consumption of material flows, energy flows and information flows in the network. Internet topology always repeatedly destroys the balance of system in self-adaptive regulation. When the network topology fluctuates between high-entropy states and low-entropy states, it is always accompanied by the birth and death of nodes and edges, requiring a steady fusion to generate a new topology. This fusion of intermediate states is called the "metabolic behavior" of the Internet. Based on the above theories, analyze and contrast the changes of standard network structure entropy of the Internet from 2011 to 2015 by taking the month as a detection unit (see Fig. 1).

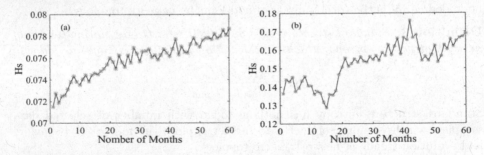

Fig. 1. Evolution of Internet standard structure entropy of the IPv4/IPv6 Internet: (a) IPv4 (b) IPv6.

It can be observed from Fig. 1 that the standard structure entropy of IPv4 and IPv6 Internet increases in time sequence, but the amplitudes are different. Within 60 months, the standard structure entropy of IPv4 Internet fluctuates from 0.071 to 0.078 nearly, entropy increment is about 0.007. In IPv6 Internet, the standard structure entropy fluctuates from 0.135 to 0.165 nearly, and entropy increment is about 0.0015, it was basically maintained at about 0.0155 in the preliminary stage. Entropy increment slightly stepped to 0.0167 only in the end of 2013. The standard structure entropy decreased significantly in 2014, and began to rise slightly in 2015. Contrast the evolution situation of IPv4 and IPv6 Internet, the entropy increment of IPv4 Internet is greater than that of IPv4 Internet. The fluctuation of Internet standard structure entropy is different under different protocols in the detection period, IPv6 is more stable than IPv4, and the entropy is IPv4 > IPv6. It is clear that the evolution of IPv6 Internet topology is more orderly, which has stronger development potential and power. The network structure entropy, as a function of describing system state, the changing trend of entropy can directly reflect the evolution direction of system state. By contrast, even for high-entropy network like IPv4 Internet, the entropy magnitude is only 10^{-2}, which is definitely not uneven network. Therefore, Internet macro topology is orderly, IPv4 Internet is still in a dynamic and unbalanced state, and the entropy increment trend observed is only a cycle occurred in the process of self-organization metabolism of network topology evolution, and network is still developing smoothly. In the future, the evolution process from IPv4 to IPv6, or other next network, IPv4 still has metabolic power, besides, will not appear to collapse topology or even end of life.

4   Self-replication Analysis of Internet Topology

The survival process of life system is trying its best to preserve its own genetic information and pass it on to the next generation through self-replication. The evolution of Internet macro topology usually emerges a kind of order, so that Internet maintains the metabolic power to fuse the new topology at all time. In order to ensure that the information will not be lost completely in the process of fusion, an inherent autocatalytic capacity required during self-organization metabolism to guide the self-combination of simple structures and maintain the topological original information. This autocatalytic capacity produces the "self-replicating behavior". Kernel number is a physical quantity that describes the hierarchical characteristics of network topology. K-core analysis is a process that deleting the nodes with small degree gradually, dissecting the network layer by layer, and mining the deep nodes and edges of the hierarchical network. The larger the kernel number is, the more central the network is. All the important information to be inherited and reserved during the evolution of the network is concentrated in the deep core structure, such the replication is meaning.

In this section, we analyze the Internet initial topology layer by layer through solving 1–25 core networks. Then, contrast the degree distribution rates of 1-core, 5-core, 9-core, 12-core, 15-core, 18-core, 21-core, 23-core and 25-core network layer by layer, observe the self-replicating behavior in each layer structure of IPv4 and IPv6 from 2011 to 2015 (see Fig. 2 and Fig. 3).

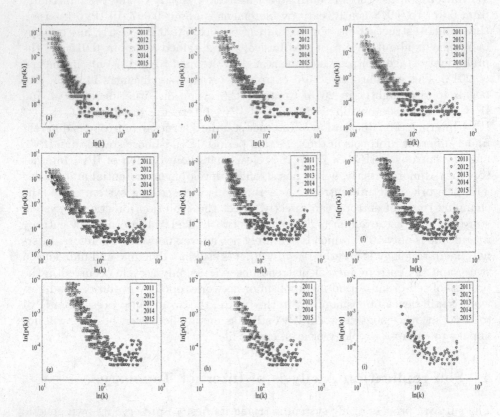

Fig. 2. Degree distribution of 1–25 core networks of IPv4 Internet topology from 2011 to 2015: (a)–(i) $1, 5, 9, 12, 15, 18, 21, 23, 25$ core network.

Figure 2 is a comparison diagram of network degree distribution rates including 1-core, 5-core, 9-core, 12-core, 15-core, 18-core, 21-core, 23-core and 25-core of IPv4 Internet topology, with double logarithmic coordinates. In different detection years, the degree distribution rate with the same core does not change much in IPv4, and is basically similar every year. Among them, the degree distribution rates of 1–10 core network topologies show linear distribution obviously, their changes are very small in five years. Throughout the detection time, the self-replicating capacity of network topology is gradually weakened in the process of core stripping of IPv4 Internet macro topology layer by layer and deep analyzing from the outer layer to the inner layer, the self-replicating behavior of outer network topology can better retain its original topology information structure.

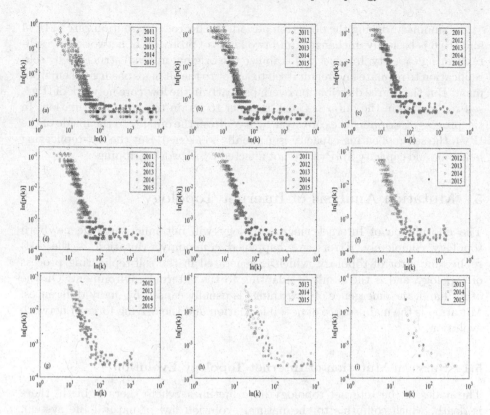

Fig. 3. Degree distribution of 1–25 core networks of IPv6 Internet topology from 2011 to 2015: (a)–(i) $1, 5, 9, 12, 15, 18, 21, 23, 25$ core network.

Figure 3 is a comparison diagram of network degree distribution rates including 1-core, 5-core, 9-core, 12-core, 15-core, 18-core, 21-core, 23-core and 25-core of IPv4 Internet topology, with double logarithmic coordinates. Throughout the different detection years, the network degree distribution rate is similar, which indicates that the self-replicating accuracy of each core network in IPv6 Internet macro topology is relatively high, but the self-replicating capacity has no hierarchical difference. But IPv6 is still a new protocol Internet, and not has the mature topology and deep core like IPv4, especially in the initial detection stage of 2011, the whole network has been stripped off when it was resolve to 18-core. During 2013–2015, the whole network can be stripped to 25-core. With the continuous stripping of the network, the nodes decrease very fast, even can be counted on one's fingers while stripping to 25-core. Since the degree distribution rate describes the distribution relationship between nodes and edges, the evolution of each core network topology has the similar characteristics and maintains the information and structure of the original network during the detection period, meanwhile, the self-replicating evolution behavior can be seen obviously in the network topology. With the rapid development of IPv6, it will evolve in the direction of complexity and deeper core in the future.

In summary, during the detection period, the degree distribution rate of IPv4 and IPv6 is basically unchanged, and two kinds of protocols of network have self-replicating capacity. Internet can stabilize the original internal structure by self-replication to regulate and repair itself to adapt to the changes of survival environment. But the degree distribution overlaps neatly in the low-core network of IPv4, self-replication in the outer network is easier to retain topology information. In the process of stripping the core layer by layer from the outside to the inside of the IPv4, the self-replicating capacity is gradually weakened. But the self-replicating behavior and capacity in IPv6 are not affected by network stripping.

5   Mutation Analysis of Internet Topology

The information of Internet macro topology will not inherit to the newborn topology completely. Errors are happened commonly. Mutation is the self-replicating error of topology, which is generated in the self-replicating process of topology, and is the result of adapting to the external environment. On the other hand, the emergence of new things is usually caused by many differences. Mutation is the main source of new information and the driving force of network evolution.

5.1   Survival Mutation of Internet Topology Evolution

The nodes in the Internet topology have lifetimes, where there is birth, there is death, which conforms to the natural evolution law of natural life system. That is a lift process from birth to dead. The lifetime of nodes can be used to analyze and predict the occurrence of network mutation. Taking the month as the detection unit, analyze the mutation effect of Internet topology in the natural evolution as a whole. The birth nodes in the current month are defined as the data that did not appear in the previous adjacent month. The dead nodes in the current month are defined as the data in last adjacent month did not appear in this month. Observe the birth rate and the dead rate of IPv4 and IPv6 Internet topology evolution in total of 60 months from 2011 to 2015 (see Fig. 4).

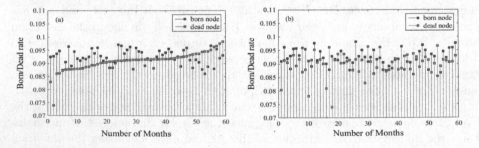

Fig. 4. Comparison of born/dead rate: (a) IPv4 (b) IPv6.

From Fig. 4, we can see that the birth rate and the dead rate in both IPv4 and IPv6 are concentrated between 7% and 10%, except for the large change of the rate in some individual month. From the view of monthly rate, it has the birth rate > the dead rate in most months, indicating that the network scale of IPv4 and IPv6 is evolving to-wards growth. The rate changes a lot in individual months, which indicate that the metabolism speed of the network itself is very fast, the birth and death of nodes have a great impact on network changes. In contrast, the emergence and lifetime of new-born nodes in IPv4 and IPv6 are basically similar, their fluctuant amplitudes are small, but the death situations of nodes are very different. Death rate of nodes in IPv4 shows a small and steady growth trend, while they fluctuate in IPv6. The differences between the two network evolutions are mainly reflected in the evolution of the dead nodes. As a new protocol network, IPv6 is still rapidly developing on optimizing its own structure, and enhancing its own viability via the topology mutation to resist environmental changes, adapting to the choice of the external environment. The development of IPv4 is relatively mature, and its survival regulation is stronger than that of IPv6. So IPv4 will not generate great differences in the metabolic and genetic process, and its whole structure will maintain a certain order. Whereas IPv6 is more likely to generate mutation when in metabolism and heredity, its structure is unstable. However, IPv6 is also evolving in survival and metabolism actively as a new generation of the Internet.

5.2   Spread Mutation of Internet Topology Evolution

In the complex network, average path length is the physical quantity that describes the speed of network information transmission. The shorter the average path is, the higher the efficiency of information transmission in the network, and the better the network performance. Analyze statistically on the average path length of the network for a total of 60 detection months from 2011 to 2015 (see Fig. 5).

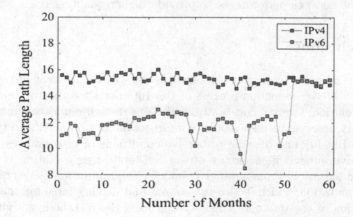

Fig. 5. Evolution of average path length.

It is obvious from the figure that the average path length in IPv4 is more relatively stable than that in IPv6, and its value is around 15. The average path length of IPv6 usually fluctuates greatly in the evolution. In the initial detecting stage, the average relative path length of the network is stable, but fluctuating actively in IPv6 from the second half of 2012 to the first half of 2015. From the average path length of the network, it has IPv4 > IPv6, which indicates that the information transmission efficiency of IPv6 is higher than that of IPv4. The high transmission efficiency puts forward a test to the performance of the whole IPv6. In a short period of time, inaccurate self-replication or erroneous data transmission makes the parameters fluctuate greatly, but not last so long for maintaining the stability of its own topology, then, the so-called mutation occurs. Therefore, the changes of the network average path length in IPv6 is more active, which indicates that IPv6 is better at adjusting in the self-replicating process, improving the network performance and information transmission efficiency, is also easier to cause transmission mutation. However, mutation is the main source of information for the Internet to generate a new topology, and is the driving force for the evolution of the system. IPv6 has more vigorous vitality to grow. According to the analysis in this section, the evolution of IPv4 and IPv6 is often accompanied by the birth and dead of nodes. Transform their own topology by adding or removing nodes and edges create error (mutation) in accurate self-replication to adjust network topology and performance. According to the analysis of the previous section, if the network standard structure entropy is IPv4 < IPv6, the macro topology order of IPv4 is better than that of IPv6. The steady evolution of transmission efficiency of IPv4 is ultimately the result of network adaptive regulation and metabolism combined with the changes of birth and dead of nodes in IPv4 and IPv6. The fluctuation amplitude of dead nodes in IPv6 is large, and the evolution of the average path length of the network is extremely active, which indicates that in the evolution process of IPv6 is more flexible and easier to adjust their own topology. It fully verifies that the routing algorithm of IPv6 updates routing tables constantly, build different logic topology, and optimize the network performance to provide higher quality service.

6   Conclusions

As an artificial complex network, Internet is regarded as a great life system to research the life reproduction process of the Internet. In order to maintain the survival evolution, Internet constantly carries out the self-replicating metabolism and heredity, reserves its own topology information and properties to the greatest extent. But because of some restrictive conditions or disturbances from the external environment, its accuracy of the self-replicating evolution is affected, resulting in mutation. From another point of view, innovation comes only when there is a mutation, which is also the source and driving force for the survival and evolution of the Internet. This paper studies the metabolic activities, self-replicating behaviors and mutational phenomenon of IPv4 and IPv6 in the process of survival and evolution, we can conclude that both IPv4 and IPv6 have vital signs and follow the law of nature. This paper realizes the research on the evolution of Internet topology inspired by biological thought and introduced by Darwin's biological evolution theory. Analyze the vital signs of metabolism, self-replication and mutation in survival evolution from the aspect of the integrity attributes of network macro topology to guide the direction of the future evolution of the Internet.

References

1. Liu, X., Wang, J.F., Jing, W., et al.: Evolution of the internet AS-level topology: from nodes and edges to components. Chin. Phys. B **27**(12), 12051 (2018)
2. Wang, J.F., Zhao, H., Liu, X.: Research on life characteristics of internet based on network motifs. arXiv e-prints, 1–13 (2016)
3. Liu, X., Zhao, H., Wang, J.F., et al.: Dissipation analysis of internet topology structure. J. Northeast. Univ. (Nat. Sci.) **36**(9), 1237–1241 (2015)
4. Zhang, S., Zhao, H.: Community identification in networks with unbalanced structure. Phys. Prev. E **85**, 066114 (2012)
5. Langton, C.G.: Artificial Lift: An Overview. MIT Press, Massachusetts (1997)
6. Ai, J., Zhao, H., Kathleen, M.C., et al.: Evolution of IPv6 Internet topology with unusual sudden changes. Chin. Phys. B **22**, 078902 (2013)
7. Kim, J.S., Goh, K.I., Salvi, G.: Fractality in complex networks: Critical and super-critical skeletons. Phys. Prev. E **75**, 016110 (2007)
8. Schneider, C.M., Kesselring, T.A., Andrade J.S.: Box-covering algorithm for fractal dimension of complex networks. arXiv preprint arXiv 1204, 3010 (2012)
9. Liu, X., Zhao, H., Li, S.F., et al.: Vital signs of IPv4/IPv6 macroscopic internet topologies. J. Comput. Res. Dev. **53**(4), 824–8331 (2016)
10. Tian, H., Zhao, H., Wang, J.F., et al.: Timing evolution and prediction of Internet transmission behavior. J. Commun. **39**(6), 116–126 (2018)
11. Zhang, Y., Yang, L., Liu, H., et al.: Applications of chaos sequence in intelligent transportation system. TELKOMNIKA **11**(9), 5210–5217 (2013)
12. Zhang, G., Quoitin, B., Zhou, S.: Phase changes in the evolution of the IPv4 and IPv6 AS-level internet topologies. Comput. Commun. **34**(5), 649–657 (2011)
13. Liu, X., Zhao, H., Zhang, J., et al.: Elastic network characteristics in internet topology. J. Northeast. Univ. (Nat. Sci.) **37**(4), 486C–495 (2016)
14. He, X., Zhao, H., Cai, W., et al.: Analyzing the structure of earthquake network by k-core decomposition. Phys. A **421**, 34–43 (2015)
15. Guo, S., Lu, Z.M., Chen, G.R: The Basic Theory of Complex Network. Science Press, Massachusetts (2012)
16. Barabsi, A.L., Albert, R.: Emerging of scaling in random networks. Science **286**, 509–512 (1999)

The Method for Image Noise Detection Based on the Amount of Knowledge Associated with Intuitionistic Fuzzy Sets

Kaihong Guo[✉] and Yongzhi Zhou

Liaoning University, Shenyang 110036, China

Abstract. The existing methods of two-stage fuzzy noise detection are all based on intuitionistic fuzzy entropy (IFE) using only the aspect of hesitation margin in the context of intuitionistic fuzzy sets (IFSs). However, there are great limitations of this type of method due to the inherent shortcomings of IFE, thus leading to limited accuracy of detecting noise. To solve this problem, we introduce in this paper the intuitionistic fuzzy knowledge measure (IFKM) into this fuzzy noise detection stage, in which the IFKM plays an important role. The fuzzification of an image at each grey level is implemented first by establishing two classes of IFSs for the foreground and background of the noisy images. The maximum amount of knowledge is then calculated, with which to determine the optimal threshold using the membership function values of the foreground and background of the noisy images, respectively. The noise pixels are detected with a noise membership function constructed by the average intensity of foreground background and each pixel value of noisy images, and removed with the iterative mean filter (IMF). By experiment, we use such performance metrics as Peak Signal-to-Noise Ratio (PSNR), Structural Similarity (SSIM), Visual Information Fidelity (VIF) and Image Enhancement Factor (IEF) to assess image quality. Furthermore, we compare denoising results of the proposed method with other state-of-the-art methods. The qualitative results show that the proposed method outperforms these algorisms. This paper applies the latest IFKM theory to the field of image noise detection for the first time, which also creates a new example for exploring the potential application areas of the theory.

Keywords: Salt-and-pepper noise · Noise detection · Image processing · Intuitionistic fuzzy knowledge measure · Noise removal · Intuitionistic fuzzy set

1 Introduction

One of the fundamental problems in image processing is noise removal. In the processing of acquisition or transmission, the images are easily corrupted by impulse noise due to shortcomings in sensors or communication channels [1]. This will cause negative effects in other image processing tasks such as image segmentation, compression, edge detection and so on. Therefore, it is necessary to reduce the noise in images before starting other image processing tasks. The main types of impulse noise are random value noise and salt

© Springer Nature Singapore Pte Ltd. 2021
Y. Gao et al. (Eds.): APWeb-WAIM 2021 Workshops, CCIS 1505, pp. 54–66, 2021.
https://doi.org/10.1007/978-981-16-8143-1_6

and pepper noise (SPN). Its characteristic is that some pixels in the image are affected and some pixels remain unaffected.

For noisy images, there are many algorithms to reduce noise, these algorithms can be roughly divided into traditional filter algorithm, fuzzy method, and other technologies based on some specific theories [2]. Most of the existing traditional filter algorithms are improved nonlinear filters, these improved nonlinear filters are mainly divided into adaptive window size and fixed window size. Adaptive window size algorithms such as Adaptive Weighted Mean Filter (AWMF) [3], Improved Adaptive Weighted Mean Filter (IAWMF) [4] and Adaptive Frequency Median Filter (AFMF) [5] and so on all use dynamic adaptive window size, and the filter window size will gradually increase before the adaptive condition is satisfied. But when the SPN ratio is high, the window size needs to be large enough, which will reduce the denoising accuracy. Fixed window size such as Modified Decision Based Unsymmetric Trimmed Median Filter (MDBUTMF) [6], Iterative Mean Filter (IMF) [7] and other algorithms. MDBUTMF has poor denoising effect when noisy image SPN ratio is high, while IMF can show excellent denoising effect in either low or high SPN ratios.

Recently, two-stage filters have been designed, which are divided into two stages: noise detection and noise removal. For example, Based-on Pixel Density Filter (BPDF) [8] and Impulse Noise Detection Technique Based-on Fuzzy Set (INDTBFS) [1], the first step of BPDF is to determine whether or not a pixel is noisy, and then decide on an adaptive window size that accepts the noisy pixel as the center. The most repetitive noiseless pixel value within the window is set as the new pixel value. INDTBFS is based on fuzzy theory and mainly relies on the intuitionistic fuzzy entropy (IFE) which is obviously shortcomings in the context of IFSs, thus leading to limited accuracy of detecting noise.

After fuzzy sets (FSs) [9] theory is proposed by Zadeh, it has been utilized in various domains especially in the field of image processing. FSs have great flexibility in dealing with uncertain information. For noisy images, vagueness arises in the selection of non-noisy pixels from a group of noisy noes using their vague brightness level and such vagueness is addressed in terms of hesitation margin. Obviously, intuitionistic fuzzy sets (IFSs) [10] are expert at expressing this vagueness. However, the existing noise detection and removal algorithms based on fuzzy theory all rely on IFE in the context of IFSs. This type of IFE only considers hesitation margin and completely ignores the inherent fuzziness between membership degrees and the non-membership degrees. This is determined by the axiomatic definition of entropy measurement. Therefore, the denoising effect of the noisy images is limited. However, Guo and Xu [11] proposed an entropy-independent axiomatic definition of intuitionistic fuzzy knowledge measure (IFKM) in the context of IFSs, which completely put aside the dependence on fuzzy entropy. In this paper, IFKM theory is introduced into the noise detection stage, and an image noise detection method based on intuitionistic fuzzy knowledge measure is proposed. The traditional nonlinear filter is organically combined with the former to noise removal.

The structure of this article is as follows. Section 2 provides an overview of knowledge measurement, Sect. 3 lists the basic knowledge needed in this article, Sect. 4 introduces the proposed noise detection method, Sect. 5 noise removal, and Sect. 6 provides the comparisons of experimental results and analysis. See Sect. 7 for the conclusion.

2   Overview of Knowledge Measurement

In order to eliminate the shortcomings of IFE, Szmidt et al. [12] first proposed the knowledge measure (KM) theory, which is related to useful information in a specific situation. Das et al. [13] and Nguyen [14, 15] believe that KM can be regarded as the negation of entropy measure in fuzzy systems. However, Guo [16] believes that entropy measure and KM have no mutual negative logical relationship, and objectively pointed out the connection and difference between entropy measure and KM. So far, in order to deal with IFKM from different perspectives, there have been many attempts. The literature [14, 15] focuses on the information content conveyed by membership degrees plus non-membership degrees. References [13, 16, 17] emphasize the inherent fuzziness of IFS/interval-valued IFSs (IVIFSs) [18]. A detailed overview can be found in [11]. In the above-mentioned exploration process of IFKM, the axiomatic definition and models mentioned all rely on entropy measures, and none of them consider the information content and inherent fuzziness at the same time. Until literature [11] Guo and Xu took the information content and inherent fuzziness into account, concurrently, and proposed an entropy-independent IFKM axiomatic definition in the context of IFSs. The axiomatic definition completely solved the shortcomings of IFE in the context of IFSs. Guo and Xu [19] further developed a bi-parametric IFKM with which to reveal some important aspects of psychological cognition hidden in the handling of IFSs. Guo and Xu [20] further explored the essence of knowledge and established a unified framework for knowledge measurement from FSs to IVIFSs, and combining normalized Hamming distance and technique for order preference by similarity to ideal solution (TOPSIS), a new model for calculating the amount of knowledge associated with IFSs is proposed. At present, the model has been successfully applied to image segmentation and decision-making under uncertainty and has achieved ideal results.

3   Preliminaries

3.1   Fuzzy Sets and Intuitionistic Fuzzy Sets

Zadeh [9] defined the notion of FSs as follows.

Definition 1 ([9]). Let X be a finite set. A fuzzy set F in X with M number of elements is defined as

$$F = \{(x, \mu_F(x))|x \in X\},$$

where the function $\mu_F(x) : X \to [0, 1]$ symbolizes the membership degree of x in X and its non-membership degree is $1 - \mu_F(x)$.

Atanassov [10] further defined the notion of IFSs below.

Definition 2 ([10]). An intuitionistic fuzzy set A on X may be mathematically computed as

$$A = \{(x, \mu_A(x), \nu_A(x)) | x \in X\},$$

where the functions $\mu_A(x), \nu_A(x):X \to [0, 1]$ respectively represent a membership degree and non-membership degree of an element x in X, with the essential condition $0 \leq \mu_A(x) + \nu_A(x) \leq 1$.

Due to lack of knowledge of whether $x \in A$ or not, there arises hesitation, which is represented by hesitation margin

$$\pi_A(x) = 1 - \mu_A(x) - \nu_A(x)$$

By utilizing this degree an IFS A can be defined as

$$A = \{(x, \mu_A(x), \nu_A(x), \pi_A(x)) | xX\}.$$

and the condition $\mu_A(x) + \nu_A(x) + \pi_A(x) = 1$ holds.

3.2 Axiomatic Definition of IFKM

Let $X = \{x_i | i = 1, 2, \ldots, n\}$ be a universe of discourse. Denoted by $IFS(X)$ the family of all IFSs in X. Let $A_i = \{x_i, \mu_A(x_i), \nu_A(x_i)\}$ be the i-th element from an IFS $A \in IFS(X), i = 1, 2, \ldots, n$. Guo and Xu [11] pointed out that there are at least two facets of knowledge associated with $A_i \in A \in IFS(X)$, one of which is called the information content while the other is the information clarity represented by $\mu_A(x_i) + \nu_A(x_i)$ and $|\mu_A(x_i) - \nu_A(x_i)|$, respectively. It can be understood that an IFS with both more information content and greater information clarity surely carries a larger amount of knowledge. With this understanding, the IFKM was axiomatized and the following-entropy-independent axiomatic definition was presented in [11].

Definition 3 ([11]). Let $A, B \in IFS(X)$. A mapping $K: IFS(X) \to [0, 1]$ is called a KM on $IFS(X)$, if K has the following properties:

$(KP_{IFS}1)K(A) = 1$ iff A is a crisp set.
$(KP_{IFS}2)K(A) = 0$ iff $\pi_A(x_i) = 1$ for $\forall x_i \in X$.
$(KP_{IFS}3)K(A) \geq K(B)$ if A has more information content and greater information clarity in comparison with B, i.e.,
$\mu_A(x_i) + \nu_A(x_i) \geq \mu_B(x_i) + \nu_B(x_i)$ and $|\mu_A(x_i) - \nu_A(x_i)| \geq |\mu_B(x_i) - \nu_B(x_i)|$, for $\forall x_i \in X$.
$(KP_{IFS}4)K(A^C) = K(A)$.

It can be seen from Definition 3 that the two facets of knowledge play a major role in measuring the amount of knowledge associated with an IFS. Clearly, the measure

K is monotonically non-decreasing with respect to the information content and the information clarity, respectively [20].

Guo and Xu [20] developed a new non-linear IFKM by using the normalized Hamming distance and the idea of the TOPSIS, which considers both information content and information clarity as mentioned before. It can be shown that $K_{IFS}(A_i) \in [0, 1]$, $i = 1, 2, \ldots, n$. For $\forall A \in IFS(X)$, it surely have [20]

$$K_{IFS}(A) = \frac{1}{n} \sum_{i=1}^{n} K_{IFS}(A_i) = \frac{1}{n} \sum_{i=1}^{n} \frac{\mu_A(x_i) + v_A(x_i)}{1 + \min\{\mu_A(x_i), v_A(x_i)\}} \quad (1)$$

3.3 Notions of REF

Definition 4 ([21]). A function $f : [0, 1][0, 1]$ is called automorphism of the unit interval. It is continuous and strictly increasing with the boundary conditions that $f(0) = 0$ and $f(1) = 1$.

Definition 5 ([21]). A function $REF : [0, 1]^2[0, 1]$ is called the restricted equivalence function, if it has the following properties:

(1) $REF(x, y) = REF(y, x) for \forall x, y[0, 1]$.
(2) $REF(x, y) = 1$ iff $x = y$.
(3) $REF(x, y) = 0$ iff $x = 1, y = 0$ or $x = 0, y = 1$.
(4) $REF(x, y) = REF(c(x), c(y))$ for $\forall x, y[0, 1]$ where c is a strong negation.
(5) $REF(x, z) \leq REF(x, y)$ and $REF(x, z) \leq REF(y, z)$ if $x \leq y \leq z$ for $\forall x, y, z[0, 1]$.

Proposition 1 ([21]). If REF is a restricted equivalence function and f an automorphism of the unit interval, then $F = f \circ REF$ is a restricted equivalence function, too.

4 Noise Detection

4.1 Image Intuitionistic Fuzzification

One of the most important problems in image denoising based on fuzzy theory is to select the appropriate membership function to achieve image fuzzification. At present, the framework based on restricted equivalence function is the most effective method for the fuzzification of the image, and is widely used [20]. In this paper, we will use the method based on restricted equivalent function to conduct noisy images intuitionistic fuzzification [21].

Let the size of the noisy grayscale with L grey levels be $M \times N$, denoted by $G = \left[g_{ij}\right]_{M \times N}$, with all pixel $0 \leq g(i, j) \leq L - 1$, for all pixel location $(i, j) \in I = \{1, 2, \ldots, M\} \times \{1, 2, \ldots, N\}$. Under the restriction of Proposition 1, constructing two classes of IFSs, L background membership degrees $\mu_{G_{B_t}}(g(i, j))$ and L foreground membership degrees $\mu_{G_{O_t}}(g(i,j))$ are constructed according to the restriction equivalent function [21], respectively

$$\mu_{G_t} = \begin{cases} F\left(\frac{g(i,j)}{L-1}, \frac{m_b(t)}{L-1}\right) = f\left(REF\left(\frac{g(i,j)}{L-1}, \frac{m_b(t)}{L-1}\right)\right) g(i,j) \leq t; \\ F\left(\frac{g(i,j)}{L-1}, \frac{m_o(t)}{L-1}\right) = f\left(REF\left(\frac{g(i,j)}{L-1}, \frac{m_o(t)}{L-1}\right)\right) g(i,j) > t; \end{cases} \quad (2)$$

$t = 0, 1, \ldots, L-1, g(i,j)G.$

where $m_b(t)$ and $m_o(t)$ represent average grey levels of background and foreground, respectively.

$$m_b(t) = \frac{\sum_{g(i,j)=0}^{t} g(i,j) \cdot h(g(i,j))}{\sum_{g(i,j)=0}^{t} h(g(i,j))}, m_o(t) = \frac{\sum_{g(i,j)=t+1}^{L-1} g(i,j) \cdot h(g(i,j))}{\sum_{g(i,j)=t+1}^{L-1} h(g(i,j))} \quad (3)$$

with $h(g(i,j))$ representing the number of occurrence of the intensity $g(i,j)$ in a given image G.

Generally, the degree of $g(i,j) \in G$ is closer to $m_b(t)$ (or $m_o(t)$), which means that the change between image intensity value and average background (or foreground) is subtle, then the value of $\mu_{G_{B_t}}(g(i,j))$ (or $\mu_{G_{O_t}}(g(i,j))$) will increase [20]. In this article, we choose $f(x) = 0.5(1 + x)$, $REF(x, y) = 1 - |x - y|^2$, $0 \leq x, y \leq 1$. In this way, Eq. 2 can be specifically expressed as

$$\mu_{G_{B_t}}(g(i,j)) = 1 - \frac{1}{2}|g(i,j) - m_b(t)|^2, \mu_{G_{O_t}}(g(i,j)) = 1 - \frac{1}{2}|g(i,j) - m_o(t)|^2 \quad (4)$$

$t = 0, 1, \ldots, L-1, g(i,j)G.$

where $m_b(t)$ and $m_o(t)$ are given by Eq. (3). $f: [0, 1] \to [0.5, 1]$ is an extension of Definition 4, which is to increase difference degree between $\mu_{G_{B_t}}(g(i,j))$ and $\mu_{G_{O_t}}(g(i,j))$ [20], the purpose is to get the best threshold for segmenting background and foreground, thereby increasing the accuracy of subsequent noise detection.

Since the subsequent calculations do not consider the non-membership degree and hesitation margin, they will not be listed anymore. For the specific construction processing, please refer to [1].

4.2 Optimal Threshold Determination and Noise Detection

Determine the Optimal Threshold. According to Eq. 1, utilizing $\mu_{G_{B_t}}(g(i,j))$ and $\mu_{G_{O_t}}(g(i,j))$ at each grey level $t(t = 0, 1, \ldots, L-1)$, calculate the amount of knowledge conveyed by two classes of IFSs, then search the grey level corresponding to the maximum amount of knowledge, then take this grey level as optimal threshold t^*. This is because in a fuzzy system, the greater the amount of knowledge, it means that the more useful information we know, the greater the certainty [20]. Corresponding to this paper, the greater the amounts of knowledge, the more correct our judgment on the classification of pixels into background or foreground, and the more accurate the subsequent noise detection.

Noise Detection. One can clearly notice that the impulse noise bears similarity with the high-frequency content of images like edges and fine details because both reflect sudden changes in pixel values [1]. The main purpose here is to use the difference degree between each pixels value with noisy images average intensity of the background, foreground to try to distinguish the pixels corrupted by SPN. For more details, please refer to [1]. According to the average intensity of foreground and background, the noise membership function is established as follows:

$$\alpha(i,j) = \begin{cases} 0, & |g(i,j) - a| < a; \\ \frac{g(i,j)-a}{b-a}, & a \le |g(i,j) - (a+b)/2| < b; \\ 1, & |g(i,j) - b| \ge b, \end{cases} \tag{5}$$

where $a = m_b(t^*)$ and $b = m_o(t^*)$ can be calculated in Eq. 3.

5 Noise Removal

5.1 IMF Denoising Algorithm

There are many methods to remove the detected the pixels corrupted by SPN. In more state-of-the-art algorithms, IMF can show excellent denoising effects in either low or high SPN ratios. IMF takes a 3×3 fixed window size, and utilizes mean value of noiseless pixel in the window to replace the center pixel [7]. Under low SPN ratio, the fixed small window size used by IMF can improve the accuracy of intensity estimation; under high SPN ratio, the iterative denoising processing ensures the thoroughness and effectiveness of denoising. Therefore, IMF algorithm is selected to process the detected noise. The output of the fuzzy filter [1] is

$$O(i,j) = \alpha(i,j) \cdot m(i,j) + (1 - \alpha(i,j)) \cdot g(i,j) \tag{6}$$

where m is the image processed by IMF.

5.2 Two-Stage Noise Detection and Removal Algorithm Based-on IFKM

The above chapters have introduced the IFKM-based noise detection and removal algorithm in detail. The following are the specific implementation steps.

Input: noisy image G:
Stage one, noise detection:

- **Step1.** G has L grey levels, with the size $M \times N$. First, at each grey level $t(t = 0, 1, \ldots, L - 1)$, utilize Eq. 3 to calculate $m_b(t)$ and $m_o(t)$, and get the background membership function $\mu_{G_{B_t}}(g(i,j))$ and foreground membership function $\mu_{G_{O_t}}(g(i,j))$ according to Eq. 4. Then according to these two values and Eq. 1 to calculate the amount of knowledge $K_t(G_t)$.
- **Step2.** Obtain the best threshold t^* according to $t^* = argmax(K_t(G_t))$.

- **Step3.** Substitute t^* into Eq. 3 to obtain $m_b(t^*)$ and $m_o(t^*)$, and then construct the noise membership function $\alpha(i,j)$ according to these two values and Eq. 5.

  Stage two, noise removal:

- **Step4.** Utilize IMF to process G to obtain m, and then remove noise according to the noise membership function $\alpha(i,j)$ and Eq. 6. Finally, the denoised image O is obtained.

  Output: denoised image O.

6 Experiment Results and Analysis

In this section, we use Peak Signal-to-Noise Ratio (PSNR) [22], Structural Similarity (SSIM) [22], Mean Structural Similarity (MSSIM) [22], Image Enhancement Factor (IEF) [23] and Visual Information (VIF) [24] to compare the denoising results of proposed method and other state-of-the-art methods (Fig. 2 and Tables 1, 2).

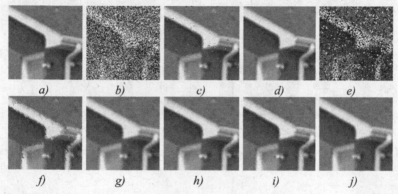

Fig. 1. Denoising results for the House image of a part of 140×140 pixel with a SPN ratio of 60%. PSNR, SSIM, VIF and IEF values of the results of the method: (a) Original image (b) Noisy image (7.35 dB, 0.0229, 0.0424, 1), (c) MDBUTMF (29.17 dB, 0.8563, 0.5130, 141.76), (d) AWMF (34.63 dB, 0.9584, 0.5179, 500.97), (e) INDTBFS (8.80 dB, 0.0486, 0.2569, 1.40), (f) BPDF (24.41 dB, 0.8204, 0.4919, 47.45), (g) IMF (36.72 dB, 0.9654, 0.5062, 864.56), (h) IAWMF (34.99 dB, 0.9619, 0.5001, 543.44), (i) AFMF (32.28 dB, 0.9401, 0.5198, 290.85), (j) Proposed method (**37.31 dB, 0.9722, 0.5077, 990.53**).

Proposed algorithm is implemented in the MATLAB2018b environment under windows 10. We use 40 images with the same size of 600×600 pixels of the TESTIMAGES dataset: https://testimages.org/ and 200 images with size of 481×321 or 321×481 pixels of BSDS dataet: http://www.eecs.berkeley.edu/Research/Projects/CS/vision/grouping/BSR/BSR_bsds500.tgz of the UC Berkeley. All images of two datasets are grayscale and are published for use under a free license.

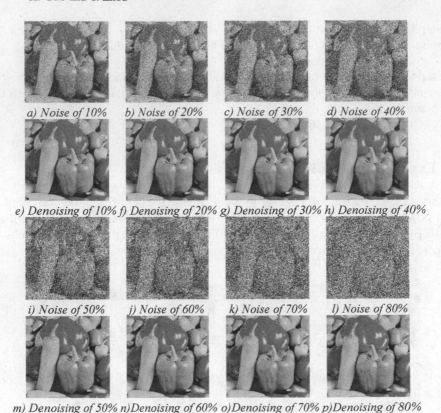

a) Noise of 10%   b) Noise of 20%   c) Noise of 30%   d) Noise of 40%

e) Denoising of 10% f) Denoising of 20% g) Denoising of 30% h) Denoising of 40%

i) Noise of 50%   j) Noise of 60%   k) Noise of 70%   l) Noise of 80%

m) Denoising of 50% n)Denoising of 60% o)Denoising of 70% p)Denoising of 80%

Fig. 2. Denoising results by proposed method for the Peppers image with the size of 512×512 pixels with different SPN ratios.

Obviously, through the experimental results, the INDTBFS algorithm based on the minimum fuzzy entropy has poor performance as the inherent shortcomings of fuzzy entropy in the context of IFSs. By observing the description of Fig. 1, the performance metrics of proposed method is better than IMF. This is because proposed method eliminates the randomness of the IMF in reducing image SPN noise through the two-stage denoising method of first detection and then removal, and noise removal is more targeted. Under the different SPN ratios of the images in above two datasets, the average VIF index of proposed method is not as good as other algorithms. A high VIF index means the higher the image quality perceived by human subjectively. However, the two criterions PSNR and SSIM are better than other algorithms under different SPN ratios, which further verify the effectiveness of proposed method in this paper.

Table 1. PSNR, MSSIM, VIF and IEF values of denoising results for the TESTIMAGES dataset with different SPN ratios.

Algorithm	Criterion	10%	30%	50%	70%	90%	MEAN
MDBUTMF	PSNR	39.01	33.48	28.83	21.75	13.66	27.47
	MSSIM	0.9776	0.9501	0.8910	0.5863	0.1619	0.7331
	VIF	0.5734	0.5105	0.4824	0.4261	0.2022	0.4481
	IEF	393.43	295.08	140.59	33.81	6.62	170.80
AWMF	PSNR	36.71	35.09	32.87	29.75	24.94	32.07
	MSSIM	0.9679	0.9538	0.9326	0.8898	0.7670	0.9090
	VIF	0.5669	0.5097	0.4870	0.4626	0.3987	0.4853
	IEF	234.03	479.60	466.22	306.02	124.13	346.80
INDTBFS	PSNR	14.14	11.03	9.26	7.42	5.69	9.42
	MSSIM	0.2524	0.1442	0.0921	0.0433	0.0131	0.1037
	VIF	0.6849	0.5252	0.4129	0.3215	0.2563	0.4345
	IEF	0.90	1.35	1.44	1.29	1.10	1.26
BPDF	PSNR	37.46	31.08	26.22	20.51	8.79	25.14
	MSSIM	0.9733	0.9266	0.8442	0.6710	0.2068	0.7481
	VIF	0.5760	0.5145	0.4788	0.3830	0.0602	0.4151
	IEF	284.04	173.83	86.88	29.48	2.37	111.82
IMF	PSNR	37.93	35.50	32.98	30.19	26.06	32.65
	MSSIM	0.9617	0.9488	0.9267	0.8860	0.7843	0.9070
	VIF	0.5716	0.5071	0.4775	0.4405	0.3526	0.4713
	IEF	275.77	509.90	494.96	363.32	175.69	388.68
IAWMF	PSNR	40.12	36.52	33.54	30.32	25.54	33.27
	MSSIM	0.9813	0.9627	0.9396	0.8989	0.7869	0.9196
	VIF	0.5728	0.5131	0.4891	0.4630	0.3962	0.4872
	IEF	513.21	654.44	540.99	350.92	142.64	458.79
AFMF	PSNR	34.16	31.00	28.50	25.97	19.84	28.16
	MSSIM	0.9571	0.9400	0.9100	0.8533	0.6510	0.8752
	VIF	0.9571	0.5089	0.4836	0.4546	0.3739	**0.5223**
	IEF	128.96	184.31	167.96	122.55	29.40	136.97
Proposed method	PSNR	41.17	36.92	33.62	30.44	26.11	**33.67**
	MSSIM	0.9835	0.9673	0.9416	0.8965	0.7888	**0.9212**
	VIF	0.5725	0.5085	0.4794	0.4426	0.3540	0.4729
	IEF	617.94	738.94	588.10	388.48	177.89	**524.53**

Table 2. PSNR, MSSIM, VIF and IEF values of denoising results for the berkeley image dataset with different SPN ratios.

Algorithm	Criterion	10%	30%	50%	70%	90%	MEAN
MDBUTMF	PSNR	37.44	31.15	27.18	21.58	14.31	26.35
	MSSIM	0.9849	0.9450	0.8631	0.5792	0.1584	0.7248
	VIF	0.5541	0.4835	0.4434	0.3785	0.2010	0.4177
	IEF	319.59	183.76	103.08	32.68	7.38	122.91
AWMF	PSNR	34.19	31.05	28.48	25.95	22.47	28.46
	MSSIM	0.9766	0.9459	0.8934	0.8167	0.6589	0.8660
	VIF	0.5462	0.4793	0.4440	0.4022	0.3160	0.4385
	IEF	131.67	168.73	145.56	108.65	59.70	128.18
INDTBFS	PSNR	14.00	11.63	10.05	8.15	6.32	9.94
	MSSIM	0.2434	0.1461	0.0970	0.0435	0.0127	0.1022
	VIF	0.6537	0.4609	0.3746	0.3230	0.3038	0.4136
	IEF	0.96	1.62	1.71	1.46	1.21	1.42
BPDF	PSNR	35.58	28.74	24.77	20.63	11.37	24.36
	MSSIM	0.9836	0.9207	0.8142	0.6355	0.2723	0.7419
	VIF	0.5554	0.4843	0.4334	0.3186	0.0524	0.3772
	IEF	180.48	93.90	57.74	29.12	4.35	69.10
IMF	PSNR	37.27	31.69	28.73	26.41	23.57	29.37
	MSSIM	0.9884	0.9525	0.8942	0.8126	0.6739	0.8705
	VIF	0.5495	0.4761	0.4334	0.3813	0.2799	0.4253
	IEF	288.39	205.54	160.70	126.43	80.82	170.73
IAWMF	PSNR	37.74	32.20	28.93	26.23	22.84	29.46
	MSSIM	0.9897	0.9598	0.9076	0.8304	0.6772	0.8804
	VIF	0.5524	0.4825	0.4459	0.4026	0.3142	0.4405
	IEF	348.51	233.38	167.70	119.40	65.91	184.09
AFMF	PSNR	31.80	29.30	26.74	24.32	19.28	26.65
	MSSIM	0.9448	0.9203	0.8661	0.7770	0.5589	0.8307
	VIF	0.5516	0.4854	0.4479	0.4036	0.3081	**0.4431**
	IEF	79.69	122.49	106.38	80.52	25.07	91.05
Proposed method	PSNR	38.70	32.46	29.12	26.58	23.61	**29.91**
	MSSIM	0.9925	0.9634	0.9094	0.8274	0.6815	**0.8817**
	VIF	0.5501	0.4769	0.4343	0.3825	0.2806	0.4262
	IEF	462.86	259.85	180.64	132.95	81.66	**215.53**

7 Conclusion

This paper proposes a knowledge-driven method for impulse noise detection and removal. It is the first time that the knowledge measure theory is applied to image impulse noise detection, and it is organically combined with the traditional nonlinear filter. Among them, knowledge measure plays a very important role, because it determines the accuracy of noise detection. This paper is another example of the application of knowledge measure in the field of image processing. The future work is to try to use the knowledge measure theory to remove other types of noise in the image.

Acknowledgements. This work is supported in part by the National Natural Science Foundation of China under Grant No. 71771110, and in part by the Planning Research Foundation of Social Science of the Ministry of Education of China under Grant No. 16YJA630014.

References

1. Ananthi, V.P., Balasubramaniam, P., Raveendran, P.: Impulse noise detection technique based on fuzzy set. IET Signal Proc. **12**(1), 12–21 (2018)
2. Zhang, F., et al.: Image denoising method based on a deep convolution neural network. IET Image Process. **12**(4), 485–493 (2018)
3. Zhang, P., Li, F.: A new adaptive weighted mean filter for removing salt-and-pepper noise. IEEE Signal Process. Lett. **21**(10), 1280–1283 (2014)
4. Erkan, U., et al.: Improved adaptive weighted mean filter for salt-and-pepper noise removal. In: 2020 International Conference on Electrical, Communication, and Computer Engineering (ICECCE), pp. 1–5 (2020)
5. Erkan, U., Enginolu, S., Dang, N., et al.: Adaptive frequency median filter for the salt and pepper denoising problem. IET Image Proc. **14**(7), 1291–1302 (2020)
6. Esakkirajan, S., Veerakumar, T., Subramanyam, A.N., PremChand, C.H.: Removal of high density salt and pepper noise through modified decision based unsymmetric trimmed median filter. IEEE Signal Process. Lett. **18**(5), 287–290 (2011)
7. Erkan, U., et al.: An iterative mean filter for image denoising. IEEE Access **7**, 167847–167859 (2019)
8. Erkan, U., Gökrem, L.: A new method based on pixel density in salt and pepper noise removal. Turkish J. Elect. Eng. Comput. Sci. **26**(1), 162–171 (2018)
9. Zadeh, L.A.: Fuzzy sets. Inf. Control **8**, 338–353 (1965)
10. Atanassov, K.: Intuitionistic fuzzy sets. Fuzzy Sets Syst. **20**(1), 87–96 (1986)
11. Guo, K., Xu, H.: Knowledge measure for intuitionistic fuzzy sets with attitude towards non-specificity. Int. J. Mach. Learn. Cybern. **10**(7), 1657–1669 (2018). https://doi.org/10.1007/s13042-018-0844-3
12. Szmidt, E., Kacprzyk, J., Bujnowski, P.: How to measure the amount of knowledge conveyed by Atanassov's intuitionistic fuzzy sets. Inform. Sci. **257**, 276–285 (2014)
13. Das, S., Dutta, B., Guha, D.: Weight computation of criteria in a decisionmaking problem by knowledge measure with intuitionistic fuzzy set and interval-valued intuitionistic fuzzy set. Soft Comput. **20**(9), 3421–3442 (2016)
14. Nguyen, N.: A new knowledge-based measure for intuitionistic fuzzy sets and its application in multiple attribute group decision making. Expert Syst. Appl. **42**(22), 8766–8774 (2015)
15. Nguyen, N.: A new interval-valued knowledge measure for interval-valued intuitionistic fuzzy sets and application in decision making. Expert Syst. Appl. **56**, 143–155 (2016)

16. Guo, K.: Knowledge measure for Atanassov's intuitionistic fuzzy sets. IEEE Trans. Fuzzy Syst. **24**(5), 1072–1078 (2016)
17. Guo, K., Zang, J.: Knowledge measure for interval-valued intuitionistic fuzzy sets and its application to decision making under uncertainty. Soft. Comput. **23**(16), 6967–6978 (2018). https://doi.org/10.1007/s00500-018-3334-3
18. Atanassov, K., Gargov, G.: Interval valued intuitionistic fuzzy sets. Fuzzy Sets Syst. **31**(3), 343–349 (1989)
19. Guo, K., Xu, H.: Preference and attitude in parameterized knowledge measure for decision making under uncertainty. Appl. Intell. **51**(10), 7484–7493 (2021). https://doi.org/10.1007/s10489-021-02317-2
20. Guo, K., Xu, H.: A unified framework for knowledge measure with application: from fuzzy sets through interval-valued intuitionistic fuzzy sets. Appl. Soft Comput. **109**(1), 107539 (2021). https://doi.org/10.1016/j.asoc.2021.107539
21. Bustince, H., Barrenechea, E., Pagola, M.: Image thresholding using restricted equivalence functions and maximizing the measures of similarity. Fuzzy Sets Syst. **158**(5), 496–516 (2007)
22. Wang, Z., Bovik, A.C., Sheikh, H.R., Simoncelli, E.P.: Image quality assessment: from error visibility to structural similarity. IEEE Trans. Image Process. **13**(4), 600–612 (2004)
23. Djurović, I.: Combination of the adaptive Kuwahara and BM3D filters for filtering mixed Gaussian and impulsive noise. SIViP **11**(4), 753–760 (2016). https://doi.org/10.1007/s11760-016-1019-x
24. Sheikh, H.R., Bovik, A.C.: Image information and visual quality. IEEE Trans. Image Process. **15**(2), 430–444 (2006)

Link Prediction of Heterogeneous Information Networks Based on Frequent Subgraph Evolution

Dong Li, Haochen Hou, Tingwei Chen$^{(\boxtimes)}$, Xiaoxue Yu, Xiaohuan Shan, and Junlu Wang

School of Information, Liaoning University, Shenyang 110036, China
dongli@lnu.edu.cn

Abstract. The problem of link prediction in heterogeneous information networks has been widely studied in recent years. It is essential to grasp the evolution law for both static information networks and dynamic information networks. However, most existing works only focus on the global topology, ignoring the impact of network microstructure on network evolution. In this paper, we present a novel link prediction method called LP-FSE (link prediction based on frequent subgraph evolution) of heterogeneous information networks. Different from traditional methods, LP-FSE makes full use of the microstructure and dynamic characteristics of networks to predict links. On one hand, frequent subgraphs are mined in heterogeneous information networks, and accord to the amount of different frequent subgraphs to predict the trend of frequent subgraphs. On the other hand, the evolution process of approximate subgraph tending to frequent subgraph is studied. Then, an approximate subgraph matching method based on matrix sliding algorithm is proposed. Experiments demonstrate the effectiveness and the efficiency of our proposed method.

Keywords: Link prediction · Heterogeneous information network · Frequent subgraph evolution · Matrix sliding

1 Introduction

In the real world, entities and relationships between entities constitute a relationship structure network, which is called an information network. An entity in information network exists in the form of node, and the relationship between nodes is called link. Link prediction is a kind of research to predict the possibility of new link between nodes. Link prediction includes prediction of unknown links and the prediction of future links. With the development of information network, link prediction has become more and more important in social relationship prediction, biological gene prediction and other fields, and link prediction has become an indispensable branch of data mining.

With the increase of information network, the categories of entities and relationships should not be ignored. The categories of entities are also diversifying, such as authors,

Y. Gao et al. (Eds.): APWeb-WAIM 2021 Workshops, CCIS 1505, pp. 67–78, 2021.
https://doi.org/10.1007/978-981-16-8143-1_7

papers and journals in the DBLP network. However, the homogeneous information network can't describe these entities and relationships at all. Compared with homogeneous information network, heterogeneous information network is more common in real world. Many scholars gradually shift their attention from homogeneous information network to heterogeneous information networks. According to the research, link prediction in heterogeneous information network has the following characteristics. Firstly, there are all kinds of nodes in heterogeneous information network, such as papers, authors, and journals in the DBLP network. Secondly, the relationship categories in heterogeneous information networks are different. For example, the relationship between author and paper is writing, the relationship between author and author is cooperation, and the relationship between paper and paper is a citation. Finally, heterogeneous information network is dynamic and sequential. The sequential characteristic is closely related to the changes of nodes and relationships, which often leads to the changes of the overall network structure.

For above characteristics of heterogeneous information network, many link prediction methods have been studied, but these methods are still shortcomings. Firstly, these methods ignore the categories of entities and relationships and only analyze them separately without considering them comprehensively. Secondly, these methods are generally only applicable to static networks, however most of heterogeneous information networks are dynamic. The prediction results at a single time are not enough to explain the rules of network structure generation and evolution. Thirdly, most of the existing methods only study single feature, either network topology or temporal feature, and do not effectively consider the combination of multiple features. Moreover, most dynamic link prediction methods only focus on global topology, ignoring the influence of network microstructure on network evolution.

According to the above shortcomings, frequent subgraph mining is considered for link prediction. As a very important network microstructure, frequent subgraph can describe the change of the whole network structure to a great extent. Moreover, we consider node categories of frequent subgraph, which can be used to predict links in heterogeneous information networks.

In this paper, we present a novel link prediction model based on frequent subgraph evolution (called LP-FSE), which makes full use of the microstructure and dynamic characteristics of networks. More specifically, we make the following contributions:

(1) We propose a frequent subgraph mining method. Different from the traditional frequent subgraph mining method, this method can effectively mine the frequent subgraphs with node category information in heterogeneous networks to use in link prediction.
(2) We propose a link prediction algorithm based on matrix sliding. According to the frequent subgraphs, approximate subgraphs of the same order that is only one link away are found, and the approximate subgraph matrix is used to slide in the whole heterogeneous information network to locate the specific nodes and links that conform to the approximate subgraph structure in the whole graph. The prediction growth of frequent subgraphs is used as the feature extraction of subgraph evolution, and the potential links between nodes are scored according to the sorting results.

(3) We conduct experimental studies based on real world data sets. Experiments demonstrate the effectiveness and the efficiency of our proposed LP-FSE method.

2 Related Work

Existing link prediction methods mainly include rely on topological features, semantic features, temporal features and mixed features.

Link prediction based on topological features considers the topological structure of the network and compares the similarity between nodes. At present, there are relatively mature link prediction methods based on topological structure such as CN (common neighbor) [1], Jaccard [2], AA [3], etc. Some works (e.g. [4–6]) consider the frequent occurrence of some special subgraphs in the network, and predict by mining these special subgraphs and finding the missing links. Compared with considering the microstructure of special subgraphs, some works consider the global structure of the network, such as Katz [7], lhn2 [8], etc. These methods of link prediction consider the relationship information in the network and calculate the similarity.

Link prediction based on semantic features is to evaluate the link probability of two nodes by calculating the similarity of their attributes. In [9, 10], the attributes and texts information of nodes are integrated into the network representation learning process, so that the learned network representation vector contains the node attribute information. In [11], node information is fused into social networks, and predicted network links by random walk method. In [12], a link prediction algorithm based on ternary closure is proposed. The algorithm calculates the weight of ternary closure of each node in the network, and uses the weight in the node similarity index to effectively improve the prediction accuracy.

Link prediction based on temporal features considers the dynamic evolution process of links and seeks temporal features from changes. Some works grasp temporal features through motif evolution and use the change probability of different temporal motifs to predict links [13, 14]. In [15], considers the directivity and historical structure of the network, and adds a time influence factor for each link to calculate. In [16], a time-series link prediction algorithm called TLP is proposed. The algorithm calculates the node representation vector at each time. Through this representation vector, we can find the change rule of nodes in the vector space. At the same time, combined with the high-order proximity characteristics between nodes, we can generate robust node vectors to maintain the network structure.

Link prediction based on mixed features is to combine multiple features to comprehensively evaluate the similarity between nodes. In [17], a link prediction method based on hierarchical proposed, which mixed graph and multi-feature fusion. In [18, 19] consider the relationship between users, event content and spatiotemporal information, and proposed an event recommendation model based on user social relationship, which effectively solved the cold start problem and can be applied to the prediction of the participation relationship between users and events.

The differences between our work and existing work are as follows:

(1) Most of the existing link prediction methods are applied to the prediction of homogeneous information networks. Although some works have studied link prediction

techniques on heterogeneous information networks, they usually treat all entities and relationships equally or analyze them independently, while ignoring the correlation between different categories of entities and relationships. The entity and relationship of different nodes are distinguished, which makes full use of the rich features of heterogeneous information network.

(2) The traditional link prediction methods only consider a single feature, because of the limited factors, it is difficult to ensure the accuracy of prediction. Although some works consider both the topological feature and the temporal feature, the combination effect is not considered, and the temporal feature is not used to mine the change of topological structure. In this paper, the temporal and topological characteristics of entities are fully considered, and the evolution process of frequent subgraphs in different time periods is considered. Moreover, the temporal characteristics are used to explore the changes of frequent subgraphs.

(3) Most of the link prediction methods based on topology structure start from the overall situation of the network, ignoring the influence of network microstructure on information network evolution. Although some works consider the use of frequent subgraphs to start from the micro, and then enlarge to the whole network, they all specify the type of frequent subgraphs artificially, which is only suitable for special networks. In the process of link prediction based on frequent subgraphs, frequent subgraphs are mined according to different networks. That is to say, the frequent subgraphs mined from different networks can effectively represent the micro characteristics of the network and have more universality.

3   Problem Definition

The problem solved in this paper can be described as follows: the temporal and undirected heterogeneous information network $G = (g1, g2, g3, \ldots, gt\ldots, gT)$ $t \in \{1,\ldots, T\}$ is composed of T time slices, and each snapshot contains the information in the same time interval of the sequential network. Given the topology of time slice $g1$ to gT, this paper proposes a link prediction method, which assigns a score to the potential links between node pairs in time slice gt. The higher the score between two nodes, the more likely the link is to be generated.

Definition 1 (Heterogeneous Information Network). For heterogeneous information networks, they use nodes to represent entities, and use edges to represent the relationship between entities. Heterogeneous information network considers the category of entities, the category of edges, attribute information of entities, the establishment time of edges, etc., which can be defined as $G = (V, E, O, R, \Phi_V, \Phi_E)$. Where, V is the set of all nodes, E is the set of edges between entities, O is the category set of nodes, and R is the category set of edges. $\Phi_V: V \to O, \Phi_E: E \to R$, which represent node category mapping function and edge type mapping function respectively. The node category of any node is $\Phi_V(v) \in O$, and the edge category of any edge is $\Phi_E(e) \in R$. Taking DBLP network as an example, the network includes different categories of entities, such as paper (P), author (A), journal (C), keyword (T) and publishing house (H). They can be represented by the relationship of writing, cooperation, quoting and publishing. The entities and relationships constitute a heterogeneous information network.

Definition 2 (Link Prediction). Given an heterogeneous information network $G = (V, E, O, R, \Phi_V, \Phi_E)$, where $V = \{v_1, v_2, v_3, \ldots, v_n\}$ represents the nodes set of the network and $E = \{e_1, e_2, e_3, \ldots, e_n\}$ represents the edges set of the network. Let the adjacency matrix of graph G be A to form a square matrix of order N, elements in matrix A show in formula 1.

$$A_{i,j} = \begin{cases} 1, & \text{if the edge } (i, j) \text{ exists in } G \\ 0, & \text{otherwise} \end{cases} \tag{1}$$

The goal of link prediction is to score the unlinked node pairs whose $A_{i,j}$ are 0 through a scoring mechanism. According to the order of score from high to low, the top-k node pairs before interception are taken as the prediction results.

Definition 3 (Frequent Subgraph). Given a single graph G, the support degree of subgraph g in a single graph G shows in formula 2.

$$Sup(g) = |g| \tag{2}$$

In other words, the support degree of a single graph is the number of times that subgraph g appears in a single graph, and calculating the support degree of a subgraph is actually testing the isomorphism of subgraphs. After the support degree is defined, the frequent subgraph in a single graph is to set a minimum support threshold (minSup) [20]. If the support degree of subgraph g is greater than or equal to this threshold, that is $Sup(g) \geq$ minSup, then the subgraph g is a frequent subgraph in the single graph.

4 Model Overview

In this section, we give an overview of our model, then explain each part of the model.

4.1 Overview of LP-FSE Model

The overall design of LP-FSE model is shown in the Fig. 1. The first step is to mine the frequent subgraphs and their number of each snapshot from the snapshot of t to $t + n$ time which is divided into n time slices in time span, and form the frequent subgraph set $M = \{t, t + 1, t + 2, \ldots, t + n\}$. According to this set, we can judge which frequent subgraph is more frequent, and the more frequent subgraph is the higher the ranking is. Then, according to the ranking, we give a different score to each frequent subgraph.

The second step is to use the frequent subgraph structure determined in the previous step as the target to find all their approximate subgraph structures. It should be considered that not all approximate subgraph structures are feasible. In this paper, we consider that a feasible approximate subgraph structure can evolve into a frequent subgraph with only one link, and it must be the same order as the frequent subgraph. Then, the matrix sliding algorithm is used to find the specific position of the approximate subgraph structure in the network at $t + n$ time, and predict whether there will become link at $t + (n + 1)$ time.

In the third step, according to the approximate subgraph structure in the actual network, judge whether the links not generated in the structure can form frequent subgraphs, and then score these links through the different frequent subgraphs, so as to achieve the purpose of link prediction.

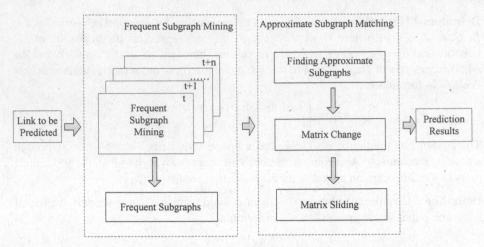

Fig. 1. Overview of LP-FSE model

4.2 Frequent Subgraph Mining

The frequent subgraphs and their numbers of each snapshot are mined from the snapshots from t to $t + n$, whose time span is divided into n time slices, to form a set of frequent subgraphs with time series $M = \{t, t + 1, t + 2,..., t + n\}$. For example, in the DBLP network, the nodes will not disappear with the change of time, and the relationship between nodes will remain forever. In other words, as long as an author has written a paper, the relationship of author node and paper node will always exist in the DBLP network. In this way, the number of times a subgraph structure appears at $t + 1$ is not less than that at t, which will lead to the evolution of frequent subgraphs in this kind of network has only an upward trend but no downward trend. At the same time, the frequent subgraphs at t are also frequent subgraphs at $t + 1$. The structure of frequent subgraphs mined at each time is described by a matrix $A = m * n$, which is called frequent subgraph number matrix, and its value is represented by $k_{t,i}$ (where t represents the time and i represents the number of frequent subgraphs). Considering that the frequent subgraph at time t is not necessarily the frequent subgraph at time $t + x$, when a certain kind of frequent subgraph does not exist at a certain time, the matrix is filled with 0. In this paper, the process of frequent subgraph mining is labeled, which can distinguish different category of nodes. Besides, the number matrix of frequent subgraphs is used to observe the evolution process of different frequent subgraphs, and then various frequent subgraphs are sorted according to the more frequent trend, and a score is given for each frequent subgraph. The higher the ranking, the higher the score, and the greater the probability of forming such frequent subgraphs.

4.3 Approximate Subgraph Matching

In this paper, when selecting approximate subgraphs of frequent subgraphs, only subgraphs of the same order are considered, and the evolution from lower order subgraphs to high order subgraphs is not considered, because it means that it is more difficult to generate links when one node has no relationship with other nodes in probability. If a complex subgraph can only be evolved from a lower order subgraph, that is, the number of edges of the frequent subgraph is less than the number of nodes, the frequent subgraph will be abandoned and no longer be used for approximate subgraph matching. At the same time, all the approximate subgraphs choose the structure which is only one link away from the frequent subgraphs, and the structure which is two or more links away from the frequent subgraphs is not considered.

In this paper, we propose matrix sliding-based approximate subgraph matching algorithm, which describes the process from searching the approximate subgraph to inputting the adjacency matrix of the approximate subgraph into the network graph matrix for sliding and matching to the specific approximate subgraph structure in the network.

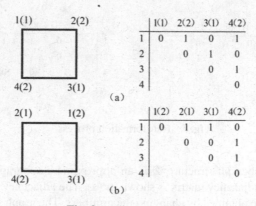

Fig. 2. Adjacency matrix

Finding Approximate Subgraphs. According to the adjacency matrix of frequent subgraphs, all approximate subgraphs with the same order difference one link of frequent subgraphs are listed and represented by adjacency matrix through continuous edge subtraction operation.

It should be considered that the adjacency matrix obtained in this step is not complete. In Fig. 2(a) and (b) are identical in structure, and the adjacency matrix is different only because of different numbers. That is to say, the same kind of approximate subgraph will have multiple adjacency matrix representations. Because matrix sliding is complete matching, if we do not get all the adjacent matrices of the same approximate subgraph structure, we will lose some qualified approximate subgraphs in the process of approximate subgraph matching. Therefore, it is necessary to transform the adjacency matrix of the approximate subgraph to make the sliding matching result more complete.

Change Matrix. Change the adjacency matrix of the approximate subgraph obtained and look for all the adjacency matrix representations of the same approximate subgraph as the input of matrix sliding.

The transformation process is show in Fig. 3.

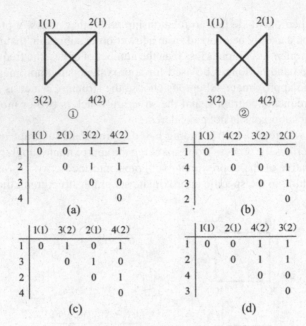

Fig. 3. Transformation process

In Fig. 3, the subgraph structure ② is an approximate subgraph of the subgraph structure ①, whose adjacency matrix is shown in (a). The adjacency matrices shown in (b), (c) and (d) can be obtained by changing the numbers. The graph structure described by these matrices is completely consistent with that of graph ②. However, it should be noted that the changed matrices may be the same, such as (a) and (d) matrices. In order to prevent the same matrix from being input into the subsequent algorithm, the same matrix will be summed and processed, and only different matrices will be recorded and input into the subsequent algorithm.

Matrix Sliding. The adjacency matrix of the approximate subgraph that needs to be slid slides in the matrix of the whole graph. The number of nodes selected during sliding is consistent with the class of the approximate subgraph. If the matrix slid out of the whole graph covers the matrix of the approximate subgraph, compare the label categories of the nodes of the approximate subgraph and the slide out matrix, and record the nodes if they are the same, otherwise they will not be recorded. Loop this process until the whole picture has been slid. These recorded nodes and matrices represent the structure and location of the subgraph exactly the same as the approximate subgraph in the whole graph, which provides input for the next link prediction.

4.4 Prediction by Adjacency Matrix

According to the approximate subgraph adjacency matrix found above, the prediction link is carried out. Links that approximate the difference between subgraph nodes will be considered as potential links. For each edges the score can be calculated according to their frequent subgraph category. It should be noted that although they are all frequent subgraphs, the number of times they appear in the network is huge. The more times they appear, the higher the probability that the approximate subgraph evolves into frequent subgraphs. In contrast, the higher the probability of generating links, the higher the score.

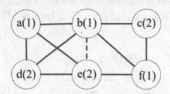

Fig. 4. Overlapping links

At the same time, considering that the edges may overlap among some subgraph structures and be predicted many times. After each loop of prediction, the score may become higher. Overlapping edges are shown in the Fig. 4. The edge between b and e is predicted twice, so the score of the edge is high. The process of prediction is shown in Algorithm 1.

Algorithm 1. Prediction Algorithm by Adjacency Matrix

Input: adjacency matrix
Output: newLinks

01. *newLinks* ← null	// initialize newLinks
02. foreach *matrix* ∈ *matrices*	// matrices obtained by matrix sliding
03. *link* ← non-existing links in matrix	// get links that do not exist in the matrix
04. for *i* ← 1 to G_N	// G_N is the amount of frequent subgraph
05. if (neighbor belongs to G_i)	
06. *score(link)*	// Calculate the score of link
07. *newLinks* ← *newLinks* + *link*	
08. end if	
09. end for	
10. end foreach	
11. return *newLinks*	

Step 1: Initialize *newlinks* (Line 1).

Step 2: For each approximate subgraph matrix and look for links that do not exist in the matrix (Line 2–3). Because the evolution in this paper only considers the formation of a link, if the frequent subgraph G_i is formed after linking an nonexistent edge, a score will be given to this edge (Line 4–8). Different frequent subgraphs have different scores. Finally, we get the score of all potential links and return this score (Line 9–11).

5 Experiment

5.1 Dataset

In this paper, we evaluate the effectiveness of our model by the dataset of DBLP (Table 1). The dataset contains the information of the authors and papers on the theory of high energy physics published from 1992 to 1995. These papers and authors form a heterogeneous information network and carry out link prediction.

Table 1. Datasets

Year	Nodes	Edges
1992	2213	4175
1993	3267	6136
1994	3821	7231
1995	3802	7489

Because of the particularity of the data set, it is not a fully connected network in the traditional sense. Most of the structures in the network are irrelevant graphs. Although there are many nodes and edges every year, because of the lack of connectivity between them, there is no difference between the prediction results of selecting too many nodes or a few nodes. In this paper, we use the combination of several subgraphs to predict. Two to three subgraphs are selected from 1992, 1993 and 1994 respectively. After mining frequent subgraphs from these subgraphs, we predict the links that are not generated in 1995 (in fact, the selected subgraph is already the one with the largest number of nodes in the dataset, about 40 nodes and more than 80 edges are selected every year).

5.2 Evaluation

In this paper uses precision and P@K as an evaluation result, the experimental results are measured. Precision is equal to the proportion of the real positive cases in the whole prediction result, which is used to measure the accuracy of the prediction result. P@K represents the accuracy of the first k predictions.

Table 2. Precision comparison of different methods

Methods	Precision
Common Neighbor (CN)	0.099
Jaccard	0.103
Random Walk (RW)	0.106
LP-FSE	0.120

We compare LP-FSE method with other traditional link prediction, as shown in Table 2, the precision of our proposed LP-FSE method is higher than other methods. Because our LP-FSE method is based on frequent subgraph evolution and focuses on the micro topology of the network, the methods selected in the comparative experiment are based on the topology. Different from some existing methods, our LP-FSE method also consider the temporal characteristics. The experimental results show that LP-FSE method is better than traditional methods, and the LP-FSE method has a wide range of application prospects. From P@K shown in Table 3, LP-FSE algorithm also has good stability. When the K is changed from 20 to 50, the prediction accuracy only decreases by 0.03, accounting for 20% of the total effect.

Table 3. Comparison on P@K

Method	P@20	P@50
LP-FSE	0.15	0.12

6 Conclusion

In this paper we propose the LP-FSE link prediction method of heterogeneous information network, which extracts frequent subgraphs and their numbers from heterogeneous network according to time sequence, and explores the evolution law of frequent subgraphs in the whole heterogeneous information network. According to the same order difference link approximate subgraph of the frequent subgraph, the matrix sliding is used to score in the whole network to complete the link prediction. The experimental results verify the advantages of this method compared with other methods. The global framework of this algorithm has nothing to do with the specific network, so it can be used in other heterogeneous information networks or homogeneous information networks after some adjustments. Next, we will further explore the application of this algorithm in other fields.

Acknowledgement. This work was supported by the Science Research Fund of Liaoning Province Education Department (LJKZ0094).

References

1. Liben-Nowell, D., Kleinberg, J.: The link-prediction problem for social networks. J. Am. Soc. Inf. Sci. Technol. **58**(7), 1019–1031 (2007)
2. Jaccard, P.: Etude comparative de la distribution florale dans une portion des alpes et des Jura. Bull. del la societe vaudoise des sciences naturelles **37**(142), 547–579 (1901)
3. Qi, G.J., Aggarwal, C.C., Huang, T.S.: Breaking the barrier to transferring link information across networks. IEEE Trans. Knowl. Data Eng. **27**(7), 1741–1753 (2015)
4. Gu, Q.Y., Ju, C.H., Wu, G.X.: Link prediction method for social networks based on subgraph evolution and improved ant colony optimization algorithm. Acta Commun. Sin. **41**(12), 21–35 (2020)

5. Zhou, P., Xiong, Y.Y.: Relationship prediction algorithm based on frequent subgraph detection in heterogeneous networks. Comput. Eng. Des. **38**(10), 2623–2630 (2017)
6. Wang, B., Pan, X., Li, Y., et al.: Road network link prediction model based on subgraph pattern. Int. J. Modern Phys. C **4**, 2050 (2020)
7. Pérez, L.G., Mata, F., Chialana, F.l.: Social network decision making with linguistic trustworthiness-based induced OWA operators. Int. J. Intell. Syst. **29**(12), 1117–1137 (2014)
8. Zhang, Q.Q.: Research on link prediction algorithm based on heterogeneous social networks. Guilin University of Technology (2018)
9. Liu, Z.M., Ma, H., Liu, S.X.: Network representation learning algorithm integrating node description attribute information. Computer application (2019)
10. Cao, R., Zhao, H.X., Ye, Z.L.: Link prediction algorithm based on network node text enhancement. Comput. Appl. Softw. **036**(003), 227–235 (2019)
11. Zhang, Y., Gao, K.N., Yu, G.: A social network link prediction method integrating node attribute information. Comput. Sci. **45**(06), 41–45 (2018)
12. Gao, Y., Zhang, Y.P., Qian, F.L.: Node similarity link prediction algorithm based on ternary closure. Comput. Sci. Explor. **11**(005), 822–832 (2017)
13. Wang, S.H., Yu, H.T., Huang, R.Y., Ma, Q.Q.: Time series link prediction method based on motif evolution. Acta automatica Sin. **42**(05), 735–745 (2016)
14. Du F, Liu Q. Link prediction method based on motif evolution in directed dynamic networks. Comput. Appl. Res. **36**(05), 1441–1445+1453 (2019)
15. Yang, R.Q., Zhang, Y.X.: A link prediction algorithm in time series directed social networks. Comput. Eng. **45**(03), 197–201 (2019)
16. Fu, H.J., Xiong, Y., Zhu, Y.Y.: TLP: a time series link prediction algorithm in dynamic networks. Comput. Eng. **46**(01), 67–73 (2020)
17. Li, D., Shen, D.R., Kou, Y., et al.: Research on a link-prediction method based on a hierarchical hybrid-feature graph. Sci. China (Inf. Sci.) **50**(2), 221–238 (2020)
18. Yin, H.Z., Zou, L., Nguyen, Q., et al.: Joint event-partner recommendation in event-based social networks. In: Proceedings of the 34th IEEE International Conference on Data Engineering (ICDE) (2018)
19. Xie, M., Yin, H.Z., Wang, H., Xu, F.J., Chen, W.T., Wang, S.: Learning graph-based POI embedding for location-based recommendation. In: Conference on Information and Knowledge Management (2016)
20. Jin, S.M., Li, Z.W., Xie, X.F., et al.: An efficient algorithm for mining frequent subgraphs of large-scale graph data. China Science and Technology Paper Online, Beijing (2016)
21. Lü, L., Jin, C.H., Zhou, T.: Similarity index based on local paths for link prediction of complex networks. Phys. Rev. E **80**, 046122 (2009)

Memory Attentive Cognitive Diagnosis for Student Performance Prediction

Congjie Liu and Xiaoguang Li[✉]

School of Information, Liaoning University, Shenyang 110036, China
xgli@lnu.edu.cn

Abstract. As the core of intelligent education, cognitive diagnosis aims to capture the proficiency of students on specific knowledge concepts. The Neural Cognitive Diagnosis (Neural CD) exploits an elegant method to simulate the interactions of student exercising process with deep neural networks. However, Neural CD still treats the student factor as static, which doesn't change after the exercising process, against the common sense that a student will gain better proficiency after practices. Furthermore, the Neural CD focuses on the current exercise despite leveraging multi fully connected layers to model the complex interactions, ignoring the relationship between exercises. In this paper, we propose the Memory Attentive Cognitive Diagnosis (MACD) for student performance prediction. Specifically, MACD introduces memory-augmented neural networks to express the student factor, which can be updated with the process of solving exercises. Moreover, MACD replaces the multi fully connected layers with multi-heads attention layers to consider the relationship between the current exercise and the past exercises. We conduct experiments on several real-world datasets and the experimental results show that our model outperforms the state-of-the-art approaches.

Keywords: Intelligent education · Knowledge tracing · Deep neural network · Cognitive diagnosis · Attention mechanism

1 Introduction

In the earlier test theory, the psychometrics-based method was item response theory (IRT) [5]. In the new generation of test theory, cognitive diagnosis is the core. Abundant works have been contributed to cognitive diagnosis, such as Deterministic Inputs, Noisy And gate model (DINA) [4], Item Response Multidimensional IRT (MIRT) [8] and Matrix Factorization (MF) [6]. Although some of these works are effective, they rely on manual-designed interaction functions which usually are simple linear arithmetical operation. Neural Cognitive Diagnosis (Neural CD) [12] incorporates neural networks to model complex non-linear interaction, preserving the explainability in the meanwhile. Despite Neural CD have proved better performance than the traditional cognitive diagnosis, the student factor is still static. Specifically, its student factor is embedded based on

Supported by the National Natural Science Foundation of China (U1811261).

the student one-hot vector, and the student's knowledge proficiency will not be updated along with the exercises. In term of common sense, a student's knowledge proficiency should change along with the process of solving exercises. Moreover, Neural CD utilizes several fully connected (FC) layers to simulate the nonlinear interaction. Compared with the traditional methods, the FC layers indeed performance better, but they still focus on the current exercise, ignoring the continuity of solving exercises and the relationship between them. When a student solves an exercise, (s)he would like to recall the memory that solving the similar exercises, and uses the experience to help solve the current exercise. It is noteworthy that the experience is not as same as the knowledge proficiency, because exercises of a certain type can contain totally different knowledge concepts.

In this paper, we address these issues in a way of proposing a Memory Attentive Cognitive Diagnosis for student performance prediction (MACD). Specifically, MACD uses multi factors such as knowledge relevancy, knowledge difficulty, and exercise discrimination to construct the exercise vector and dynamically traces the student' s knowledge proficiency in term of DKVMN. Different from the static student factor of vanilla Neural CD, MACD utilizes the memory networks to generate the student factor which can be updated along with student's path of solving exercises. In addition, MACD leverages the multi-head attention networks to simulate the interaction process between the student and the exercise. Neural CD uses FC layers to augment the non-linearity in interaction, but ignores the continuity of solving exercises. A student's exercises are usually correlated in order to teach the student step by step. Based on this, we exploit the attention mechanism to trace the relationship between exercises, simulating the recalling process in the solving process. Our experiments show that MACD outperforms other baseline methods, and make more accurate predictions.

Our main contributions are summarized as follows:

1. Our model proposes a DKVMN-based method to dynamically update the student factor making the interaction function closer to the real learning process.
2. Our model simulates the process of recalling historical exercises by taking exercises' relationship into account using attention mechanism.
3. Experiments on real-world online datasets prove that MACD outperforms other baseline methods.

2 Related Works

Cognitive Diagnosis. Item response theory (IRT) [5] was the main psychometrics-based method of the earlier test theory of intelligent systems. It modeled the result of a student answering an exercise as the interaction between the trait features of the student ability (θ) and the exercise difficulty (β). They modeled the interaction in a logistic way, e.g., sigmoid($\alpha(\theta - \beta)$), where α is the exercise discrimination parameter. In DINA [4], θ and β were binary, and β was gotten from the human labeled exercise-knowledge correlation matrix which is known as Q-matrix. Besides, guessing and slipping parameters were also taken into consideration to simulate the interaction. After that, researchers added extra parameters in IRT, and latent trait was extended to multidimensional (MIRT) [8]. Wang et al. [12] proposed Neural Cognitive Diagnosis (Neural CD) framework by introducing neural networks into cognitive diagnosis in order to model the complex non-linear interactions. By combining the neural networks and cognitive diagnosis, Neural CD achieved both high accuracy and interpretation. The framework of neural CD is shown in Fig. 1. Although Neural CD outperforms the traditional cognitive diagnosis, its student factor remains unchangeable as the former. And Neural CD only considers the non-linearity in interactions, but ignores the relationship between the current exercise and past exercises which can represent the student's historical experience in solving exercises.

Deep Learning Based Knowledge Tracing. Knowledge Tracing (KT) is another main type of approaches of intelligent education system. The earliest method was BKT [3]. With the rise of deep learning, deep knowledge tracing (DKT) [7] was proposed. DKT based on recurrent neural networks (RNN), exploits the utility of latent states in RNN to learn student's knowledge proficiency. Zhang et al. proposed dynamic key-value memory networks (DKVMN) [13] to trace students' proficiency of concepts by introducing memory-augmented neural networks (MANN) [9]. Abdelrahman et al. [1] used attentive mechanism, Sun et al. [10] added behavior features of students and Ai et al. [2] considered the containment relationship between concepts, they all improved DKVMN from different sides. Although these DKVMN-based methods simulate the process by which students get proficiency of concepts with elegant methods, they are all issued with the lack of interpretability.

As illustrated above, both cognitive diagnosis and KT methods have their own advantages and disadvantages. Towards this end, in this paper, we propose a Memory Attentive Cognitive Diagnosis for student performance prediction which combines the neural cognitive diagnosis and DKVMN to dynamically update the student concept proficiency, then uses attention mechanism to model the interaction process of a student's solving an exercise.

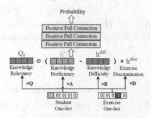

Fig. 1. Neural cognitive diagnosis model.

3 Model

We first formally describe the overview of MACD model. After that, the details of DKVMN-Neural CD layer and Attention layer will be introduced respectively.

3.1 Model Overview

MACD defines knowledge concept set as $C = \{q_1, q_2, \ldots, q_{|C|}\}$; problem set as $P = \{p_1, p_2, \ldots, p_{|P|}\}$. Students are independent of each other. One student's interaction at timestamp t is defined as (q_t, p_t, r_t), where $q_t \in \mathbb{N}^+$ is concept index, $p_t \in \mathbb{N}^+$ is problem index and $r_t \in \{0, 1\}$ is the student's response. $r_t = 1$ means the student responded correctly and $r_t = 0$ means the wrong response. Given a student's historical interaction records $S = [(q_1, p_1, r_1), \cdots, (q_{t-1}, p_{t-1}, r_{t-1})]$, the goal of MACD is to predict the student's response r_t to exercise (q_t, p_t).

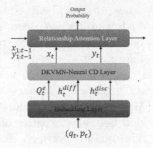

Fig. 2. Overview of MACD.

As shown in Fig. 2, MACD first inputs tuple (q_t, p_t) into the embedding layer to construct three factors: knowledge relevancy, knowledge difficulty, exercise discrimination. In DKVMN-Neural CD layer, MACD traces the student's knowledge proficiency using DKVMN and makes up the interaction function using the four factors. After that, MACD transmits the exercise vector and the interaction vector to the relationship attention layer for tracing the relationship of exercises. The attention layer simulates the interaction process that the student solves the exercise and recalls the historical experience. Finally, the layer outputs the predicted response to the target exercise.

3.2 Embedding Layer

In embedding layer, the input is a student's interaction tuple (q_i, p_i, r_i) at any timestamp $i \in (1, t)$. MACD embeds the exercise's concept q_i with embedding matrix $\boldsymbol{E}^{diff} \in \mathbb{R}^{d \times |C|}$, acquiring the concept difficulty factor $\boldsymbol{h}_i^{diff} \in \mathbb{R}^d$. For exercise problem p_i, MACD embeds it with the exercise discrimination matrix $\boldsymbol{E}^{disc} \in \mathbb{R}^{|P|}$, obtaining the exercise discrimination factor $h_i^{disc} \in \mathbb{R}$. The knowledge relevancy factor is about concept and exercise both. MACD embeds them with two different matrices $E^c \in \mathbb{R}^{|C|}$ and $E^e \in \mathbb{R}^{|P|}$ respectively, and gets their vectors $c_i \in \mathbb{R}^d$ and $b_i \in \mathbb{R}^d$. Then the knowledge relevancy factor is gotten by a sigmoid-activated fully connect layer as follows:

$$Q_i^e = \text{Sigmoid}(W_q(c_i \circ b_i) + b_q) \tag{1}$$

where $\circ$ is element-wise product. Finally, MACD embeds response r_i with $E^r \in \mathbb{R}^{d \times 2}$ and gets the response vector $g_i \in \mathbb{R}^d$. Different from the vanilla DKVMN exercise vector Q which is gotten from q_i directly [13], MACD constructs the exercise vector $x_i \in \mathbb{R}^d$ referring to the factors mentioned above.

$$x_i = Q_i^e \circ h_i^{diff} \times h_i^{disc} \tag{2}$$

After constructing x_i, MACD traces the student factor using DKVMN by reading and writing key-value memory networks.

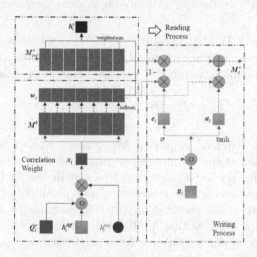

Fig. 3. The architecture for DKVMN-Neural CD layer. The model is drawn at the timestamp i.

3.3 DKVMN-Neural CD Layer

Correlation Weights and Reading Process. For the cognitive diagnosis, the key is to grasp the student's proficiency. The architexture of DKVMN-Neural CD layer is shown in Fig. 3. In order to trace the student proficiency factor h_i^s at timestamp i, firstly, MMAKT needs to know the correlation weights of the exercise x_i on the knowledge concept key matrix M^k. The correlation weight is computed by taking the softmax activation of the inner product of x_i and each key slot M_j^k.

$$w_{i,j} = \text{Softmax}(x_i{}^T M_j^k) \tag{3}$$

where $M_j^k \in \mathbb{R}^d$ is the jth slot of the knowledge concept key matrix $M^k \in \mathbb{R}^{|C| \times d}$, and $w_{i,j}$ is the jth value of $w_i \in \mathbb{R}^{|C|}$ representing the correlation weight between the exercise and each latent concept.

Then MACD takes the weighted sum of M_{i-1}^v with respect to w_i, and gets the student proficiency factor at timestamp i, defined as $h_i^s \in \mathbb{R}^d$.

$$h_i^s = \sum_{j=1}^{|C|} w_{i,j} M_{i-1,j}^v \tag{4}$$

where $M_{i-1,j}^v \in \mathbb{R}^d$ is the jth slot in the knowledge concept value matrix $M_{i-1}^v \in \mathbb{R}^{|C| \times d}$.

In term of Neural CD, MACD constructs the interaction vector $y_i \in \mathbb{R}^d$ with the knowledge concept relevancy factor Q_i^e, the concept difficulty factor h_i^{diff}, the exercise discrimination factor h_i^{disc} and the student proficiency factor h_i^s.

$$y_i = Q_i^e \circ (h_i^s - h_i^{diff}) \times h_i^{disc} \tag{5}$$

Writing Process. After h_i^s is gotten, M_{i-1}^v needs to be updated to M_i^v for the next computation at timestamp $i+1$ according to the correctness of the student's answer. Unlike the knowledge growth vector in vanilla DKVMN, MACD constructs the update vector with the element-wise product between the exercise vector x_i and the response vector g_i in order to match with x_i. The following update process is the same as that of DKVMN, including erase subprocess and add subprocess. The eraser vector $e_i \in \mathbb{R}^d$ is calculated as follows:

$$e_i = \text{Sigmoid}(W_e(x_i \circ g_i) + b_e) \tag{6}$$

where $W_e \in \mathbb{R}^{d \times d}$, $b_e \in \mathbb{R}^d$ are parameter vectors. Similarly, the add vector $a_i \in \mathbb{R}^d$ is calculated as follow:

$$a_i = \text{Tanh}(W_a(x_i \circ g_i) + b_a) \tag{7}$$

where $W_a \in \mathbb{R}^{d \times d}$, $b_a \in \mathbb{R}^d$ are parameter vectors. Then, M_{i-1}^v is updated to M_i^v with e_i and a_i:

$$M_{i,j}^v = M_{i-1,j}^v(1 - w_{i,j}e_i + w_{i,j}a_i) \tag{8}$$

where $\mathbf{1}$ is d dimension one vector.

3.4 Attention Layer

The interaction process is modeled by several fully connected layers in Neural CD, but it only focuses on the current exercise, forgetting a student usually requires the experience of solving the similar exercises when answering an exercise. To simulate the recalling process, MACD uses multi-head attention networks [11] to trace exercises' relationship, and makes prediction with them. We provide the explanation of the multi-head attention networks in the next subsection.

Multi-head Attention Networks. The multi-head attention networks inputs Q_{in}, K_{in}, and V_{in} which represent the sequence of queries, keys, and values respectively. And the networks apply the single self-attentive mechanism *heads* times for the same one input sequence in different projection matrices. By the dot-product between the query and the key, the network gets the relevance of the value to the corresponding query. To avoid the influence on the current position from the future information, the network utilizes a masking mechanism which replaces upper triangular part of the product matrix with $-\infty$ in order to zero out the attention weights of the subsequent positions. As shown in Fig. 4, the single self-attention head is,

$$head_i = \text{Softmax}(\text{Mask}(\frac{Q_{in}W_i^Q(K_{in}W_i^K)^T}{\sqrt{d}}))V_{in}W_i^V \tag{9}$$

where W_i^Q, W_i^K, W_i^V are projection matrices.

The final output of the multi-head attention networks is the concatenated tensor of heads attention heads multiplied by W^O.

$$\text{MultiHead}(Q_{in}, K_{in}, V_{in}) = \text{Concat}(head_1, ..., head_{heads})W^O \tag{10}$$

Feed-Forward Networks. From above we can see that the multi-head attention networks are linear. To add non-linearity to the model, position-wise feed-forward networks are applied.

$$\text{FFN}(M_H) = \text{ReLU}(M_H W_1^{FF} + b_1^{FF})W_2^{FF} + b_2^{FF} \tag{11}$$

where $M_H = \text{Multihead}(Q_{in}, K_{in}, V_{in})$ and $W_1^{FF}, W_2^{FF}, b_1^{FF}$ and b_2^{FF} are weight matrices and bias vectors.

Relationship Attention Layer. To model the recalling process, MACD utilizes multi-head attention networks to trace the relationship between the current exercise and the past exercises. The attention layer inputs the target exercise x_t as query, the exercise sequence $x_{1:t}$ as keys, the interaction sequence $y_{1:t}$ as values, and outputs the prediction vector f_t:

$$\hat{f}_t = \text{MultiHead}(Q_{in} = x_t, K_{in} = x_{1:t}, V_{in} = y_{1:t}) \tag{12}$$

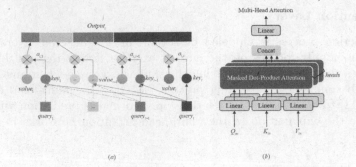

Fig. 4. (a) Network of a masked dot-product attention. At each timestamp the attention weights are estimated for each of the previous element only. When jth element is query and i element is key, attention weight is a_{ij}. (b) Network of a multi-head attention. Each multi-head attention consists of several attention layers running in parallel.

$$\boldsymbol{f}_t = \text{FFN}(\hat{\boldsymbol{f}}_t) \tag{13}$$

In relationship attention layer, the relationship between exercises is represented by the attention between query and key, the historical experience is the value, and the output vector $\boldsymbol{f}_t$ is seen as the gathering of the experience of the past and the present. Then MACD converts $\boldsymbol{f}_t$ to the predicted response o_t via several sigmoid activated fully connected layers like the neural CD.

$$\boldsymbol{f}_t^1 = \text{sigmoid}(\boldsymbol{f}_t \boldsymbol{W}_{f1} + \boldsymbol{b}_{f1}) \tag{14}$$

$$\boldsymbol{f}_t^2 = \text{sigmoid}(\boldsymbol{f}_t^1 \boldsymbol{W}_{f2} + \boldsymbol{b}_{f2}) \tag{15}$$

$$o_t = \text{sigmoid}(\boldsymbol{f}_t^2 \boldsymbol{W}_{f3} + \boldsymbol{b}_{f3}) \tag{16}$$

Here we need to restrict each element of $\boldsymbol{W}_{f1}, \boldsymbol{W}_{f2}, \boldsymbol{W}_{f3}$ to be positive in order to satisfy the monotonicity assumption [12].

Finally, all the trainable parameters in MACD can be learned by minimizing the cross-entropy loss between o_t and r_t.

$$\mathcal{L} = -\sum_{i \in |S|} (r_t \log(o_t) + (1 - r_t) \log(1 - o_t)) \tag{17}$$

4 Experiments

4.1 Datasets and Evaluation Metric

We use two real-world datasets to evaluate MACD. The ASSIST2009 dataset was collected from an online tutoring platform. The slepemapy.cz dataset was collected from a geo-online education system.

ASSISTments2009: Dataset contains 4151 students, 110 concepts, 16891 problems and 325637 interactions.

slepemapy.cz: Dataset contains 87952 students, 1458 concepts, 10087305 interactions, and every problem has one concept.

We use the Area Under Curve (AUC) and the Accuracy (ACC) as evaluation metrics. AUC is defined as the area under the receiver operating characteristics curve, representing the predictive performance of the model. Higher AUC and ACC values indicate better performance of the model.

4.2 Baseline Methods and Approach

We compare MACD against the baseline methods: DKVMN and Neural CD. The parameters are set as follows:

DKVMN: Hyperparameters are set following Zhang et al. [13]. The memory size is 50, and memory dimension is 200. The learning rate is 0.001 with Adam optimizer.

Neural CD: Hyperparameters are set following Wang et al. [12]. The learning rate is 0.001 with Adam optimizer.

MACD: For different two datasets, the memory sizes of M^k and M^v are set according to the number of concepts in each dataset. The batch size is 30, and the learning rate is 0.01 with Adam optimizer. Hyperparameter d is determined by comparing AUC values. The results of testing are shown in Table 1. We can find in the table that the AUC values are higher than the others when $d = 24$. It can be seen that when d is set too low, the performance of the model decreases; when d is set too high, there are too many parameters in the model, which easily lead to overfitting. So, d should be chosen according to the result of experiments.

Table 1. AUC results with different d

ASSIST2009		slepemapy.cz	
d	AUC	d	AUC
16	0.750 1	16	0.817 1
24	**0.764 4**	**24**	**0.818 6**
32	0.751 9	32	0.817 9

4.3 Results and Discussion

In this paper, for each dataset, 20% learners are used as the test set, 60% are used as the training set and 20% are used as the validation set to adjust hyperparameters and early stop.

Table 2 shows the results of AUCs and ACCs of MACD and the other two baselines on two datasets. The results show that MACD generally outperforms the baselines. DKVMN only uses concepts of exercises to trace the knowledge proficiency, and abandons the other information in exercise such as discrimination or knowledge relevancy. On the contrary, Neural CD makes full use of exercise information but doesn't dynamically update the student proficiency.

MACD balances the advantages of the two models, achieving to update the proficiency while preserving the explainability. In addition to that, MACD also simulates the experience-recalling process with the attention mechanism which can lead to better performance.

Table 2. Experimental results on student performance prediction

Model	ASSIST2009		slepemapy.cz	
	AUC	ACC	AUC	ACC
DKVMN	0.745 9	**0.726 8**	0.785 7	0.803 8
Neural CD	0.723 7	0.705 9	0.793 1	0.796 7
MACD	**0.764 4**	0.722 0	**0.821 5**	**0.814 5**
MACD-N	0.750 0	0.718 7	0.818 6	0.809 1

In addition, we also conduct experiments denoted as MACD-N without the attention mechanism. The results show the performance of MACD-N is slightly lower than MACD, but still higher than the other models.

Figure 5 shows the AUC plots of MACD and the other two methods on the ASSIST2009's validation set in 50 epochs. The results show that Neural CD has obviously overfitting problem. DKVMN performs well in preventing overfitting, but its AUC performance is inferior to that of MACD because it only makes predictions based on the student's knowledge states without taking into account the other information in exercises and historical information. Compared with the two baselines, MACD shows excellent results on predicting student's performance.

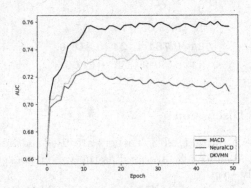

Fig. 5. Performance on ASSIST2009 validation set.

4.4   Visualizing Parameters

We choose a student in ASSIST2009 randomly, and visualize learning records over a period time. In Fig. 6, the x-axis represents the student's interaction records, where q_t is exercise's concept, and r_t is student's response in (q_t, r_t). The y-axis represents four concepts which these records contain. The data is the visualized student knowledge proficiency h^s. From the data we can see that after the student's correct answer about concept 31 at timestamp 4, h^s improved correspondingly, and at timestamp 11, h^s declined after the incorrect answer. This indicates that MACD can dynamically update students' knowledge proficiency according to students' answers.

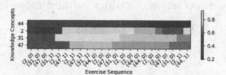

Fig. 6. Knowledge proficiency output result of MACD.

In the attention layer we take the attention weights of the first 10 exercises to make illustration. In Fig. 7, the x-axis represents the past exercises, and the y-axis represents the target exercises. The data is the visualized attention weight. Taking the exercise 7 as example, the weights of the exercise 3 and 5 are significantly higher than others. That means the exercise 3 and 5 are more similar with the exercise 7 than the other exercises. So, when the student solves the exercise 7, (s)he would like to recall the situation that (s)he solved the exercise 3 and 5 to help solve the exercise 7. With the attention layer, MACD can capture the historical experience in the interaction process.

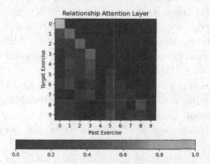

Fig. 7. Attention weights in relationship attention layer.

5   Conclusions and Future Works

In this paper, we utilize memory networks to enhance Neural CD and improve the dynamic character of the student factor. And we leverage attention mechanism to trace the relationship between exercises in order to model the recalling process in solving exercises. The experiments show that our model outperforms DKVMN and Neural CD.

In future research, we will explore the following aspects:

1. For cognitive diagnosis, we consider using other reasonable methods to construct other factors, such as discrimination factor or knowledge relevancy factor.
2. For the interaction process, we consider adding more factors, such as time interval, etc.

References

1. Abdelrahman, G., Wang, Q.: Knowledge tracing with sequential key-value memory networks. In: Proceedings of the 42nd International ACM SIGIR Conference on Research and Development in Information Retrieval, pp. 175–184 (2019)
2. Ai, F., et al.: Concept-aware deep knowledge tracing and exercise recommendation in an online learning system. International Educational Data Mining Society (2019)
3. Corbett, A.T., Anderson, J.R.: Knowledge tracing: modeling the acquisition of procedural knowledge. User Model. User-Adap. Inter. 4(4), 253–278 (1994). https://doi.org/10.1007/BF01099821
4. De La Torre, J.: Dina model and parameter estimation: a didactic. J. Educ. Behav. Stat. 34(1), 115–130 (2009)
5. Embretson, S.E., Reise, S.P.: Item Response Theory. Psychology Press (2013)
6. Koren, Y., Bell, R., Volinsky, C.: Matrix factorization techniques for recommender systems. Computer 42(8), 30–37 (2009)
7. Piech, C., et al.: Deep knowledge tracing. arXiv preprint arXiv:1506.05908 (2015)
8. Reckase, M.D.: Multidimensional item response theory models. In: Reckase, M.D. (ed.) Multidimensional Item Response Theory. SSBS, pp. 79–112. Springer, New York (2009). https://doi.org/10.1007/978-0-387-89976-3_4
9. Santoro, A., Bartunov, S., Botvinick, M., Wierstra, D., Lillicrap, T.: Meta-learning with memory-augmented neural networks. In: International Conference on Machine Learning, pp. 1842–1850. PMLR (2016)
10. Sun, X., Zhao, X., Ma, Y., Yuan, X., He, F., Feng, J.: Muti-behavior features based knowledge tracking using decision tree improved DKVMN. In: Proceedings of the ACM Turing Celebration Conference-China, pp. 1–6 (2019)
11. Vaswani, A., et al.: Attention is all you need. arXiv preprint arXiv:1706.03762 (2017)
12. Wang, F., et al.: Neural cognitive diagnosis for intelligent education systems. In: Proceedings of the AAAI Conference on Artificial Intelligence, vol. 34, pp. 6153–6161 (2020)
13. Zhang, J., Shi, X., King, I., Yeung, D.Y.: Dynamic key-value memory networks for knowledge tracing. In: Proceedings of the 26th International Conference on World Wide Web, pp. 765–774 (2017)

The Second International Workshop on Deep Learning in Large-scale Unstructured Data Analytics

A Study on the Privacy Threat Analysis
of PHI-Code

Dongyue Cui and Yanji Piao[✉]

Yanbian University, Yanji, Jilin, China
piaoyanji@ybu.edu.cn

Abstract. During the period of epidemic prevention and control, personal health information code (PHI-code) is widely used in China and has brought significant benefits, including helping to manage the flow of people effectively and to slow down the spread of corona virus. However, the personal information security issues involved in the use of PHI-code has also attracted people's attention. To solve this problem, we can use the security threat modeling to strategize and protect personal information. There are various security threat modeling techniques exist today, such as LINDDUN, PASTA and NIST. In this paper, we analyze all possible privacy threats contained in PHI-code by using LINDDUN, which is a threat modeling technique for personal information protection and is not used in the analysis of domestic PHI-code. Finally, we derive threat mitigation strategies based on the security evaluation results. This study will improve the future public health information collection system towards more privacy protection and promote the stable development of society.

Keywords: PHI-code · Privacy threats · Mitigation strategies

1 Introduction

In December 2019, a newly emerged virus received widespread attention all over the world [10]. On February 11, 2020, WHO announced that the novel disease was to be named "corona virus disease-2019 (COVID-19)". As of 11 June 2021, there are more than 110,000 confirmed cases in 34 provinces (regions) of China and more than 174,900,000 confirmed cases in 214 countries [5]. In order to identify the movements of COVID-19 infected patients and find close contacts of them, countries around the world have developed different methods, such as 'Exposure Notification' [14], 'ROBERT' [3], 'COVIDSafe' [4] and 'PHI-code' [13]. In this paper, we only consider PHI-code which is widely and effectively used in various regions of China. Although PHI-code plays an important role in the prevention and control of the epidemic, the privacy issues it brings cannot be ignored. First of all, the PHI-code involves important personal information such as name, telephone number and certificate number. The background uses traffic data such as civil aviation and railway to track user's position and path. These private information is easy to be obtained by attackers, and the disclosure of information may bring great inconvenience to our daily lives. Then, PHI-code cannot be separated from

© Springer Nature Singapore Pte Ltd. 2021
Y. Gao et al. (Eds.): APWeb-WAIM 2021 Workshops, CCIS 1505, pp. 93–104, 2021.
https://doi.org/10.1007/978-981-16-8143-1_9

information sharing, which often involves the health status and other issues. Once the patient's information, such as travel trajectory, work unit, home address, etc., is released by malicious attackers, it will cause widespread concern in society and may lead to discrimination against him, causing serious negative effects.

In this paper, we do the security evaluation of PHI-code by applying LINDDUN and derive response measures. First, in order to analyze the system structure, we create data flow diagram (DFD) of PHI-code in Sect. 3.1. Second, we apply LINDUN to identify security threats in more detail in Sect. 3.2. Third, we identify the threat scenarios in Sect. 3.3. Fourth, we derive response measures in Sect. 4.

2   Related Work

2.1   PHI-Code

Personal health information code (PHI-code) is an information-based electronic certificate used to prevent and control the spread of COVID-19, which can also be called health QR code. The QR code generated by mobile application can be used as a certificate to entry and exit in public places in China [15]. A reference model of PHI-code [13] is shown in Fig. 1 to illustrate PHI-code system that will be discussed throughout this paper. The system is used by four actors: PHI-code scanning application, PHI-code displaying application, personal health information service, PHI-code service. *PHI-code scanning application* can identify personal health information to keep track of user's health. *PHI-code displaying application* can provide personal information to primary-level organization such as personal health information, travel path, COVID-19 test results, etc. *Personal health information service* is a controller of the personal information. *PHI-code service* can generate and verify health code displayed to primary-level organization.

The process of PHI-code system is as follows:

1) A user who wants to display PHI-code sends PHI-code request to PHI-code service by using PHI-code displaying application.
2) PHI-code service returns PHI-code to the user.
3) The user shows PHI-code to primary-level organization who installs PHI-code scanning application.
4) The primary-level organization sends query request to personal health information service.
5) Personal health information service sends request to PHI-code service in order to verify the PHI-code.
6) PHI-code service returns back the query index or other information according to verification results.
7) Personal health information service searches the health information and sends it to the primary-level organization.

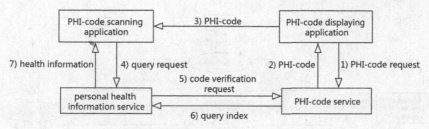

Fig. 1. PHI-code reference model.

2.2 Security Threat Modeling

Adam Shostack [1] described that the security threat modeling identifies security risks and threats of the product during software development life-cycle. Kim Wuyts and Wouter Joosen [16] presented a comprehensive framework to model privacy threats in social network 2.0 application. In the paper [11], the authors analyzed the privacy threats of smart home hub by using LINDDUN and presented evaluation criteria against corresponding security requirements. Hojun Lee and Sangjin Lee [8] performed the security assessment of "Exposure Notification" [14] to derive all possible threats. They presented common evaluation criteria applicable to all types of contact tracing technologies [9] as well. In the paper [6], the authors built target privacy threat model for "Exposure Notification" system and outlined threat mitigation strategies. Some researchers studied other systems' security assessment in order to protect privacy [2, 7, 17].

In this paper, we apply security threat modeling techniques to derive all possible threats of PHI-code system. Firstly, in order to analyze the structure systematically, we use a data flow diagram (DFD) to represent the PHI-code system. We can learn all the components of the system and data flow from DFD. Secondly, LINDDUN [12], one of the threat modeling method, is applied for identifying security threats that may exist in the PHI-code system. LINDDUN is a representative method from the perspective of privacy. Each letter of "LINDDUN" stands for a privacy threat type [12]: L (Linkability), I (Identifiability), N (Non-repudiation), D (Detectability), D (Information Disclosure), U (Content Unawareness), N (Policy and consent Noncompliance).

We study whether security risks and threats exist in the PHI-code system through DFD and LINDDUN, and thus we can derive all types of risks and threats that may have negative impacts within PHI-code system.

3 Security Evaluation of PHI-Code

This section provides security evaluation to identify privacy threats in th PHI-code system. In Sect. 3.1, DFD is created. In Sect. 3.2, privacy threats are mapped to the DFD. In Sect. 3.3, threat scenarios are identified.

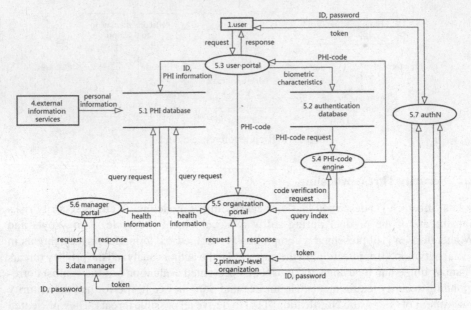

Fig. 2. DFD representing the PHI-code system.

3.1 Creating the Data Flow Diagram (DFD)

The PHI-code system is graphically represented by using a data flow diagram, which has following 4 major types of building blocks: external entities, data stores, processes and data flows. The DFD is shown in Fig. 2 to illustrate PHI-code system. There are 4 external entities, 2 data stores, 5 processes and 24 data flows. Table 1 shows a detailed description of DFD elements.

3.2 Mapping Privacy Threats to DFD

LINDDUN. LINDDUN is a systemic approach for privacy threat modeling. The following section describes 7 main privacy threat components that are handled in LINDDUN:

- Linkability: Being able to sufficiently distinguish whether two or more items of interest are related or not within the system.
- Identifiability: Being able to identify the subject within the anonymity set.
- Non-repudiation: An attacker can gather evidence to counter the claims of the repudiating party.
- Detectability: An attacker can sufficiently distinguish whether an item of interest exists or not.
- Information Disclosure: Being unaware of the consequences of sharing data. Exposing personal information to someone who are not authorized to see it.

Table 1. Detailed description of the DFD element of PHI-code.

Group	Component	Explanation
Entity	User (1)	A user who installs PHI-code displaying application
	Primary-level organization (2)	A user who installs PHI-code scanning application. ex., hospital, railway station, airport, etc.
	Data manager (3)	Personal data controller
	External information services (4)	transport department, telecommunications department, etc
Data store	PHI database (5.1)	Storage for personal data (ex.:health information, itinerary information, etc.)
	Authentication database (5.2)	Storage for authentication data
Process	User portal (5.3)	The user interface front end
	PHI-code engine (5.4)	Generate and verify PHI-code
	Organization portal (5.5)	The organization interface front end
	Manager portal (5.6)	The manager interface front end
	AuthN (5.7)	The common interface for entities authorization
Data flow	User - portal (1 - 5.3)	Data flow that requests login
	Portal - user (5.3 - 1)	Data flow that responses login
	Primary-level organization - organization portal (2 - 5.5)	Data flow that requests login
	Organization portal - primary-level organization (5.5 - 2)	Data flow that responses login
	User portal - PHI database (5.3 - 5.1)	Data flow that updates personal health information
	User portal - authentication database (5.3 - 5.2)	Data flow that passes identification information of biometric characteristics
	Authentication database - PHI-code engine (5.2 - 5.4)	Data flow that requests PHI-code
	External information services - PHI database (4 - 5.1)	Data flow that updates personal extra information
	PHI-code engine - user portal (5.4 - 5.3)	Data flow that sends PHI-code
	User portal - organization portal (5.3 - 5.5)	Data flow that displays PHI-code

(*continued*)

Table 1. (*continued*)

Group	Component	Explanation
	Organization portal - PHI-code engine (5.5 - 5.4)	Data flow that requests PHI-code verification
	PHI-code engine - organization portal (5.4 - 5.5)	Data flow that provides query index
	Organization portal - PHI database (5.5 - 5.1)	Data flow that sends query request
	PHI database - organization portal (5.1 - 5.5)	Data flow that returns health information
	Data manager - manager portal (3 - 5.6)	Data flow that requests login
	Manager portal - data manager (5.6 - 3)	Data flow that responses login
	Manager portal - PHI database (5.6 - 5.1)	Data flow that sends query request
	PHI database - manager portal (5.1 - 5.6)	Data flow that returns health information
	User - authN (1 - 5.7)	Data flow that inputs ID and password
	AuthN - user (5.7 - 1)	Data flow that responses token
	Primary-level organization - authN (2 - 5.7)	Data flow that inputs ID and password
	AuthN - primary-level organization (5.7 - 2)	Data flow that responses token
	Data manager - authN (3 - 5.7)	Data flow that inputs ID and password
	AuthN - data manager (5.7 - 3)	Data flow that responses token

– Content Unawareness: Not understanding the information disclosed to the system.
– Policy and consent Noncompliance: Not following the privacy policies or the existing user consents.

Mapping Privacy Threats to DFD. As shown in Table 2, LINDDUN provides a template that shows which threat components are applicable to each DFD element. It depicts the relationship between LINDDUN privacy threat components and DFD elements types. In Table 2, 'X' means that the DFD element type in the corresponding column is susceptible to the privacy threat type of the selected row. For example, a data store is subject to linkability, identifiability, non-repudiation, detectability, non-compliance and information disclosure.

Table 2. Available DFD elements per LINDDUN.

	L	I	N	D	D	U	N
Entity	X	X				X	
Data store	X	X	X	X	X		X
Process	X	X	X	X	X		X
Data flow	X	X	X	X	X		X

We make some assumptions to reduce the number of 'X' and simplify the analysis process. In Table 3, we identify the privacy threat types by following the template showed in Table 2 based on our assumptions. 'X' represents a potential privacy threat at a corresponding DFD element in the system which we actually consider. We omit relevant description of Table 3 and replace it with threat tree in Sect. 3.3.

Assumptions are as follows:

1) We believe that the back-end of the health code built by the government is capable of resisting outsider threats, so internal processes will only be affected by insider threats. Because one process threat can represent all of them, we combine all process threats into one and examine only one. In a same manner, we combine all data flow threats into one and examine only one.

2) Non-repudiation threats do not exist in the system, as the data stores, processes and data flow. For example, there are abundant evidence from external information services (transport department) whether a person has been to a high-risk area.

3) Detectability is not considered a threat for this specific system. The privacy concerns of the PHI-code system are focused on the data itself, not on the detectability of it.

4) Identifiability of entities (user, primary-level organization, data manager and external information services) is not considered a threat, when all entities use the health code system, they need to login with real-name authentication. For the health of the entire society, the identity of the entity cannot be hidden.

5) Identifiability of the data flows is not considered a threat, as entities other than user need to identify people who use the health code system.

6) Linkability of entities (user, primary-level organization, data manager and external information services) is not considered a threat, as user needs to login with real-name authentication, it is not able to be replaced by a pseudonym.

7) Linkability and identifiability do not apply to data flows, as knowing who send a "request" does not infringe the privacy. The personal privacy is invaded when the content of the information is revealed.

8) Non-compliance is an important threat that it is associated with a threat for a whole system.

9) Content unawareness only applies to an entity such as the user and the external PHI services, who input additional information in the system, as the primary-level organization does not input any data, and the data manager do not add any information.

Table 3. LINDDUN for PHI-code system.

	Threat target	L	I	N	D	D	U	N
Entity	User (1)						X	
	Primary-level organization (2)							
	Data manager (3)							
	External information services (4)						X	
Data store	PHI database (5.1)							
	Authentication database (5.2)							
Process	User portal (5.3)							X
	PHI-code engine (5.4)							X
	Organization portal (5.5)							X
	Manager portal (5.6)							X
	AuthN (5.7)							X
Data flow	User - portal (1 - 5.3)							X
	Portal - user (5.3 - 1)							X
	Primary-level organization - organization portal (2 - 5.5)							X
	Organization portal - primary-level organization (5.5 - 2)							X
	User portal - PHI database (5.3 - 5.1)							X
	User portal - authentication database (5.3 - 5.2)							X
	Authentication database - PHI-code engine (5.2 - 5.4)							X
	External information services - PHI database (4 - 5.1)							X
	PHI-code engine - user portal (5.4 - 5.3)							X
	User portal - organization portal (5.3 - 5.5)							X
	Organization portal - PHI-code engine (5.5 - 5.4)							X
	PHI-code engine - organization portal (5.4 - 5.5)							X
	Organization portal - PHI database (5.5 - 5.1)							X

(continued)

Table 3. (*continued*)

Threat target	L	I	N	D	D	U	N
PHI database - organization portal (5.1 - 5.5)							X
Data manager - manager portal (3 - 5.6)							X
Manager portal - data manager (5.6 - 3)							X
Manager portal - PHI database (5.6 - 5.1)							X
PHI database - manager portal (5.1 - 5.6)							X
User - authN (1 - 5.7)							X
AuthN - user (5.7 - 1)							X
Primary-level organization - authN (2 - 5.7)							X
AuthN - primary-level organization (5.7 - 2)							X
Data manager - authN (3 - 5.7)							X
AuthN - data manager (5.7 - 3)							X

10) We assume that the data stores are fully protected in PHI-code system and that any attacks are not possible for malicous purpose.
11) Information Disclosure threat of the data flows is not considered as they are highly-unlikely to occur because they take a lot of analysis and the extracted information is not in correspondence of the effort.
12) Information Disclosure threat of the processes is not considered as we assume all the processes are implemented correctly.

3.3 Identifying Threat Scenarios

Threat tree is the result of the threats that may occur in LINDDUN threat elements confirmed in the previous section. Table 4 reflects the threat tree for PHI-code system. We map the specific threats of this case to the LINDDUN PRIVACY THREAT TREE CATALOG [12]. Then we use misuse cases to elaborate on specific threats as shown in Table 5. The misuse case shows how threats occur through scenario and result. The misuse case structure is provided in LINDDUN PRIVACY THREAT MODELING [16].

Table 4. Threat tree for PHI-code.

Unawareness of entity			
1	U		
	1.1	U_2: Unaware of stored data	
Non-compliance			
1	NC		
	1.1	NC_2: Incorrect or insufficient privacy policies	
		1.1.1	NC_3: Inconsistent/insufficient policy management

Table 5. Misuse cases of threat tree.

MUC	Description
MUC01	Threat Tree: U (Unawareness) Summary: The data subjects (user and external information services) are not aware of the situation where personal information is stored Primary misactor: Management Basic path: Bf1. User and external information services cannot determine what personal health information has been saved Consequence: User and external information services cannot identify personal (health) information even if it is revealed DFD element(s): 1 user 4 external information services
MUC02	Threat Tree: NC (Non-Compliance) Summary: The data subjects (primary-level organization,data manager) do not process personal (health) information in compliance with legislations, policies or user agreement Primary misactor: Insider/outer person/system operator Basic path: Bf1. The misactor fails to comply with the country's policy or legislation (e.g. the user's personal (health) information is revealed to third parties) Consequence: The user's personal (health) information is shared to others and even spreaded widely in society, causing inconvenience and psychological pressure on user's live. When detected, the reputation of the health sector even the entire government is damaged DFD element(s): process Data flow

4 Threat Mitigation Strategy

According to the analysis of this paper, the main threats to PHI-code are as follows:

1) Unawareness of user and external information services
2) Non-Compliance of data flow and process

We propose mitigation strategies to cope with the privacy threat in this section. Table 6 represents the privacy requirements and mitigation strategies corresponding to the misuse cases. Regarding misuse case 1, the mitigation strategy we give is "review data", which refers to the use of feedback tools to improve users' privacy awareness, such as the service agreement and privacy policy that are prompted when the user uses the PHI-code for the first time to explain in detail which user's data will be collected and stored. Regarding misuse case 2, the mitigation strategy we give is "perfect policies", which contains two-phase policies, development phase and application phase. First of all, during the application development phase, developers should design system in compliance with perfect policies for privacy and data protection. More specifically, developers can hire employee who is responsible for making the policies compliant or hire external company for compliancy auditing. Secondly, in the stage of using the PHI-code system to collect information and conduct contact tracing, the policies should clearly specify what relevant personnel who can access user information can and cannot do. For example, whoever discloses users' information will be severely penalized. The use of GDPR and GITHUB is also a way we can improve policy and consent compliance.

Table 6. Mitigation strategy.

Misuse cases	Privacy requirements	Mitigation strategies
Content Unawareness (MUC01)	Data subject are aware of what data the system has actually stored and collected	Review data
Policy and consent Noncompliance (MUC02)	Each company has own privacy policies. Users should be able to know who has access to their information and what happens with their data	Perfect policies

5 Conclusion and Future Work

It is essential to use PHI-code in China no matter where people go so as to prevent and control major public health emergencies. As PHI-code is used more and more frequently, privacy issues are getting more and more attention. In order to better protect personal information, it is necessary to analyze the security threats of PHI-code. In this paper, we analyze threats contained in PHI-code by using LINDDUN, which is a threat modeling for privacy protection.

To begin with, we create DFD to understand the structure of PHI-code system. Then, we use LINDDUN framework to build a privacy threat model which reflects all possible privacy threats faced by the PHI-code system. Furthermore, we map the threat tree and provide misuse case based on template and our assumptions. Finally, we provide mitigation strategies that should be addressed the threats defined in our model.

Further research can be undertaken in the following areas: First, we will further apply LINDDUN to analyze COVID-19 tracing systems from different countries. Second, we will further analyze privacy threats for making a comparison among countries.

Acknowledgements. This work was supported by Application Foundation Program of the Science and Technology Development Plan of Yanbian University of China (NO.602021014).

References

1. Adam, S.: Threat Modeling: Designing for Security. Wiley, New York (2014)
2. Cartagena, A., Rimmer, G., Dalsen, T.V., Watkins, L., Rubin, A.: Privacy violating open-source intelligence threat evaluation framework: a security assessment framework for critical infrastructure owners. In: 2020 10th Annual Computing and Communication Workshop and Conference (2020).
3. Cla, C., et al.: ROBERT: ROBust and privacy-presERving proximity Tracing, HAL-Inria hal-02611265 (2020)
4. COVIDSafe. http://www.springer.com/lncs. Accessed 29 Jun 2021
5. Epidemic Map. https://ncov.dxy.cn/ncovh5/view/pneumonia_peopleapp?from=timeline&isappinstalled=0. Accessed 11 Jun 2021.
6. Gangavarapu, A., Daw, E., Singh, A., Iyer, R., Raskar, R.: Target privacy threat modeling for covid-19 exposure notification systems (2020). https://arxiv.org/abs/2009.13300
7. Gholami, A., Lind, A.S., Reichel, J., Litton, J.E., Edlund, A., Laure, E.: Privacy threat modeling for emerging BiobankClouds. Proc. Comput. Sci. **37**, 489–496 (2014)
8. Hojun, L., Sangjin, L.: A study on the security evaluations and countermeasure of exposure notification technology for privacy-preserving COVID-19 contact traving. J. Korea Ins. Inf. Secur. Cryptol. **30**(5), 929–943 (2020)
9. Hojun, L., Sangjin, L.: Evaluation criteria for COVID-19 contact tracing technology and security analysis. J. Korea Ins. Inf. Secur. Cryptol. **30**(6), 1151–1166 (2020)
10. Hsia, W.: Emerging new coronavirus infection in Wuhan, China: situation in early 2020. Case Stud. Case Rep. **10**(1), 8–9 (2020)
11. Jaehyeon, P., Sooyoung, K., Seungjoo, K.: Study of security requirement of smart home hub through threat modeling analysis and common criteria. J. Korea Ins. Inf. Secur. Cryptol. **28**(2), 513–528 (2018)
12. LINDDUN: Threat Tree Catalog. https://linddun.org/linddun-threat-catalog. Accessed 11 Jun 2021
13. Personal health information code-Reference Model GB/T 38961–2020. http://openstd.samr.gov.cn/bzgk/gb/
14. Privacy-Preserving Contact Tracing-Apple and Google. (2020). https://covid19.apple.com/contacttracing
15. Shukla, M., Rajan, M.A., Lodha, S., Shroff, G., Raskar, R.: Privacy guidelines for contact tracing applications. arXiv e-prints (2020)
16. Wuyts, K., Joosen, W.: LINDDUN privacy threat modeling: a tutorial. Report CW 685 (2015)
17. Wuyts, K., Scandariato, R., Joosen, W.: Empirical evaluation of a privacy-focused threat modeling methodology. J. Syst. Softw. **96**, 122–138 (2014)

Logistics Policy Evaluation Model Based on Text Mining

Guangwei Miao[1], Shuaiqi Wang[2], and Chengyou Cui[1(✉)]

[1] Yanbian University, Yanji, Jilin, China
cycui@ybu.edu.cn
[2] Jilin University, Changchun, Jilin, China

Abstract. With the development of computer technology and Internet technology, people can get a large amount of data directly through the Internet. This paper builds a model of policy evaluation on text mining by using text mining techniques, and builds policy evaluation indicators through word breaker technology, word frequency statistics, construction thesaurus, subject analysis, and co-existing analysis. Using 10 logistics policy texts in Heilongjiang Province, 12 logistics policy texts in Jilin Province and 8 logistics policy articles in Liaoning Province between 2015 and 2020, a total of 30 logistics policy texts were used as the research objects of this paper, and then the PMC model was used to score the logistics policy text, the PMC surface was drawn to visualize the focus of logistics policy documents in different provinces, and the policy evaluation models constructed by experts were combined with the policy interpretation documents to verify that the policy evaluation models built by text mining technology were effective.

Keywords: PMC · Text excavation · Policy evaluation

1 Introduction

With the development of computer technology and Internet technology, people can get a large amount of data directly. Under the application of computer technology, text mining based on big data has gradually become an essential field in data science research in natural science and social science. At present, the primary sources of data for text mining are papers, public text files, books, periodicals, web pages, social media user reviews, etc. In the study of policy evaluation, empirical analysis is mainly used, yet some evaluation framework models and evaluation methods have their own defects, such as the use of gray correlation model, the optimal straightness of the specified indicators is too subjective, may have a relatively large error situation; The PMC (Policy Modeling Consistency Index) model used in this paper solves the above problems, in the PMC model first of all, the use of text mining technology to determine indicators, in order to reduce subjectivity and improve the accuracy of the set indicators, the use of text mining technology, not only can quickly and accurately find out the critical words set up by the text, but also by changing different text mining statistical algorithms to achieve

© Springer Nature Singapore Pte Ltd. 2021
Y. Gao et al. (Eds.): APWeb-WAIM 2021 Workshops, CCIS 1505, pp. 105–116, 2021.
https://doi.org/10.1007/978-981-16-8143-1_10

different focus of research, At the same time, combined with text mining technology relative to the previous policy interpretation, combined with text mining technology can simultaneously deal with a large number of text policy, for the improvement of efficiency is a leap.

Policy tools are policy measures and means adopted by the government to achieve specific policy objectives, which embody the policy maker's governing philosophy and policy value [1]. It is an effective and vital way to study public policy to analyze the structure based on policy text and to combine and construct policy into a series of essential unit elements, such as human resources and capital and information. Because of the classification of policy tools, academic circles propose different dividing criteria based on multiple perspectives. Howlett and others have designed three types of policy tools: mandatory, resource-based and hybrid, based on the intensity of government intervention and the degree of market autonomy [2]. McDonnell and others generalize policy tools into motivational, command-, learning- and system-based based policies based on context, purpose, and effectiveness [3]. The most recognized and widely used are the Rothwell and Zegveld classification methods, which divide policy tools into supply-oriented, environmental, and demand-based according to the level of policy's role in science and technology activities [4]. This classification method will integrate the complex policy system to carry out precise and effective structured treatment and three policy tools to refine further and explain. For improving the efficiency of government industrial governance, innovation-driven transformation plays an important role [5]. Policy evaluation refers to the functional activities of the whole policy system and process using scientific evaluation criteria and methods [6].

This paper obtains the data of this paper. It constructs the evaluation index in the PMC model by using the PMC model combined with text mining technology. Finally, it uses the logistics policy text of the three eastern provinces to apply the method.

2 Methodology

2.1 Model Framework

The required analytical data text and put it in a set database. Compared with the previous traditional method, manually go through the relevant information, and then enter the information has many disadvantages, this stage requires a lot of time and human resources, and with the increase of text, the input errors will be more and more, and manual operation efficiency is very low. However, using reptile methods can avoid the above problems to ensure the speed of text acquisition and ensure the accuracy of the information obtained (Fig. 1).

2.2 Text Crawl

The structure of the network reptiles showing the various components and their interdependencies, as shown in Fig. 2. The URL manager is responsible for managing the URL addresses that you are about to access, including the set of URL addresses to be accessed and already visited, the web downloader is responsible for downloading HTML

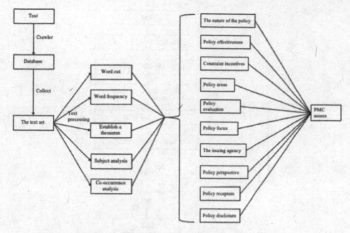

Fig. 1. A model of policy evaluation based on text mining

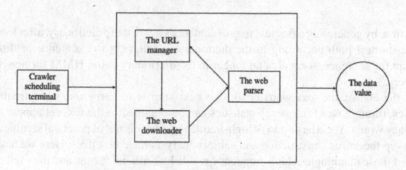

Fig. 2. Web crawler structure diagram

data in the URL page, and the web parser is primarily responsible for interpreting the information in the HTML and passing the new URL to the URL manager.

A workflow diagram of a web crawler, as shown in Fig. 3. First, select a portion of the seed URL seat reptile to start data, choose a good URL placed in the geometry with crawling, select the URL to be crawled, access download, and do de-reprocessing, repeated execution know the conditions met after exiting the loop.

2.3 Text Mining Statistics

Because the language used in the text is Chinese, Chinese compared to English has the characteristics of continuity, unlike the English kind of natural separation through spaces, though connected, sometimes multiple words constitute a word, sometimes a word is a word, so the difficulty of word separation, in this paper is mainly using the python language jieba library for word-sharing processing, jieba word-sharing has three sub-word patterns, namely: precise mode, full mode, search engine mode. The algorithms for all three modes are different, and here we use precise patterns to avoid over-repetition, which increases the complexity of subsequent studies. Jieba word breaker precision

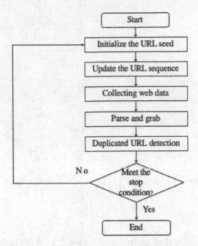

Fig. 3. Web crawler work flowchart

mode, first, by generating a ringless map of sentences against the dictionary, after looking for the shortest path according to the dictionary to intercept the sentence or directly intercept the sentence, for words no longer in the dictionary using HMM for new word discovery.

At the end of the text word break, the next step is to carry out word frequency statistics, through word frequency statistics can be obtained in the text set appear high-frequency words. Yet, a large part of high-frequency words is not of practical significance. Follow-up thesaurus construction and subject analysis have no effect. Here we actively filter out those meaningless high-frequency words by artificial means and then select the remaining words into our corpus.

Word Frequency Analysis is a statistical and analytical analysis of the number of appearances of essential words in the body of a document and an essential means of text mining. It is a traditional and representative content analysis method in literature econometrics. The basic principle is to determine the hot spot and its change trend by how frequently the word changes.

After building the thesaurus, reuse python to return to our text set to get the word cloud for the text set. Word cloud refers to the visual prominence of keywords that appear more frequently in mass text content. The higher the frequency of use, the higher the frequency of occurrence in the article. At the same time, the co-existing analysis, "common present", refers to the phenomenon of the information described in the characteristics in the literature, the relevant characteristics of the co-present quantitative study to reveal the content of the information and the knowledge implied by the characteristics. In the text set, the common occurrence of words in the thesaurus selected by the Institute can determine the direct relationship between those words in the text set, general understanding, the more times appear at the same time, indicating that the relationship between the two keywords is closer, the more worthy of our study. Thus, the frequency of the occurrence of the subject words in the statistical text set to form a common network composed of these words through the network between the nodes and the thickness of the connection can reflect the affinity of these words. Finally, we completed the construction of our evaluation index system.

2.4 PMC Model

This paper uses the PMC model, which was proposed in 2010 by Ruiz Estrada Etal who argued that everything is a movement and that no one related variable should be overlooked when building a policy model. In evaluating a policy, the model emphasizes that there is no limit to the number and weight of variables, so that the advantages and disadvantages and internal consistency of a policy can be analyzed from multiple dimensions, which is the more advanced policy evaluation method in the world. The specific steps for constructing the PMC model include identifying variable classification parameters, building a multi-input output table, measuring, and calculating the PMC index, and generating PMC curved surfaces. In the text, the model contains a total of 9 main variables, each of which contains several sub-variates, using binary 0 and 1 to balance all sub-variates, the eligible assignment is 1, the non-qualified assignment is 0, and then the mean of the sub-variate is the final score of each main variable, and the score of the main variable is summed together to get the PMC index value. Rank evaluation criteria based on PMC index score. The PMC exponential model calculates the PMC score based on the numerical values quantified by text mining technology and makes PMC curved surface diagrams to show the advantages and disadvantages of policy text in each dimension, which greatly avoids the disadvantages of the limited research dimension of other policy evaluation methods (Fig. 4).

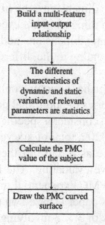

Fig. 4. PMC analysis framework

In this paper, through the method of determining variables proposed by Ruiz Estrada, combined with the application of Zhang Yong An's PMC index model to the policy evaluation index system [7], combined with the characteristics of the logistics policy text, this paper finally establishes nine primary variables and 29 secondary variables.

After determining the secondary index and judgment criteria of the PMC evaluation index of logistics policy, the variables excavated by text are quantified into our evaluation index system through the formula (1) (2) (3) (4) to get a score for each text.

$$Xnn \sim N[0, 1] \tag{1}$$

$$X = \{XR : [0, 1]\} \tag{2}$$

$$X_t(\sum_{j=1}^{n} \frac{X_{tj}}{T(X_{tj})})t = 0, 1, 2, 3 \ldots \ldots \infty \tag{3}$$

$$PMC = X_1(\sum_{j=1}^{5} \frac{X_{1j}}{T(X_{1j})}) + \cdots + X_9(\sum_{j=1}^{1} \frac{X_{9j}}{T(X_{9j})}) \tag{4}$$

As shown in Table 1, give the scoring criteria for our policy evaluation. Based on the above calculations, our multi-input output table can be constructed, and the PMC index of our policies can be calculated by replacing the values of the multi-input output table with formula (4).

Table 1. Score evaluation form

PMC score	9–8	7.99–6	5.99–4	3.99–0
Evaluation	Excellent	Well	Common	Badness

By calculating each of the nine dimensions, the highest score for each dimension is one point, the variables under each dimension are evenly distributed to one point under the modified dimension, and the nine dimensions add up to a total of nine points, with different scoring intervals corresponding to the different ratings of the file, as shown in Table 1.

The introduction of policies often has a certain continuity. The emergence of a subject policy will be accompanied by a series of additional policies. The evaluation of policies should look at a specific policy and consider a specific policy system. This paper selects the East Three Provinces 2015–2020 a total of six years of all logistics policies as the object of research, to the province as a unit, all the province's policy documents as a system as a complete quantitative evaluation. The PMC curved surface can directly give the calculation results of the PMC exponential model in different dimensions. The curved surface construction of this paper is based on the PMC index of text. The PMC curved surface drawing is a matrix of 3×3 involving nine variables.

3 Empirical Analysis

3.1 The Subject of the Study

The text crawled a total of 30 logistics policy texts from the government websites of Liaoning, Jilin and Heilongjiang provinces from 2015–2020, eight from Liaoning Province, ten from Heilongjiang Province and twelve from Jilin Province.

3.2 Object Research

By summarizing all the policy text, using python's jieba toolkit for this article word segmentation, through word segmentation and word frequency statistics, you can quickly extract the core content highlighted in the policy document. Due to space issues, this article selects the section as the presentation, as shown in Table 2. Build a logistics policy thesaurus by removing words that don't make much sense and leaving behind words that make sense for research.

Table 2. Top words

Word	Frequency	Word	Frequency	Word	Frequency
Logistics	890	Construction	571	Traffic	366
Enterprise	782	Server	418	Distribution	360
Develop	645	Manage	402	Safe	359
Delivery	620	Job	367	Charge	352

Compared to unwanted words, we need particularly limited words, so we need words into the thesaurus so that in the statistics, will only count the words we need, significantly reduce the complexity of analysis, but also increase the accuracy of analysis, improve the scientific and accuracy of text analysis but also for the follow-up word cloud map and network map analysis to lay the foundation for thematic analysis and the establishment of a common network to provide data.

Using the thesaurus filtering, we can quickly gather and reflect the core and focus of all current policy texts, as shown in Table 3. We visualize these high-frequency theme words to see more intuitively what our policy text highlights and direction, as shown in Fig. 5. As can be seen from the word cloud map, the current analysis of the policy text, the main emphasis is on enterprises as the main policy body, by development, construction, management, support, express delivery, cities, and other words reflect the most emphasized content in the policy text, through express delivery can be drawn, the current logistics policy text, emphasizing more is the logistics industry express plate. Focus on express delivery development and planning.

Table 3. The thesaurus words

Word	Frequency	Word	Frequency	Word	Frequency
Enterprise	782	Serve	418	City	326
Develop	645	Manage	402	Encourage	270
Delivery	620	Traffic	366	Backing	234
Construction	571	Distribution	360	Nudge	220

Fig. 5. Word cloud

A co-existing network diagram that the central location of demand indicates that it is associated with many other words, and when a word is associated with many words, it appears in the central location, as shown in Fig. 6.

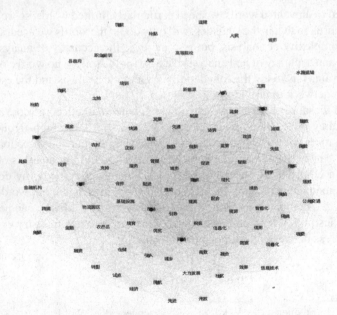

Fig. 6. A common network

In this paper, the PMC evaluation index system is constructed by the results obtained by text mining and Zhang Yong An's research results and by combining the characteristics of logistics policy.

The index system framework proposed in this paper is based on the framework of the innovative policy evaluation model, combined with text mining technology to find out the characteristics of the subject of logistics policy text to build a logistics policy text policy evaluation system.

An index evaluation system with nine primary variables and twenty-nine secondary variables is obtained based on the above process, as shown in Fig. 7.

The first-level variable	The secondary variable	The secondary variable	The secondary variable	The secondary variable	The secondary variable
The nature of the policy	Forecast	Suggestion	Steer	Supervise	
Policy effectiveness	Long-term	Medium-term	Short-term		
Policy areas	Economy	Society	Technology	Politics	Environment
Policy perspective	Macro	Medium-view	Microcosmic		
Policy subject	Government				
Policy receptors	Government	Enterprise	Public		
Policy focus	Scientific and technological innovation	System construction	Market guidance	Results promotion	
Incentive constraints	Talent introduction	Capital investment	Tax benefits	Equity incentives	Subsidy
Policy disclosure	Openness				

Fig. 7. Policy evaluation system

3.3 Results and Analysis

Based on the score calculation of the policy documents of Poly evaluation system and formula (4), we list the text with the highest and lowest scores for logistics policies in Heilongjiang, Jilin and Liaoning provinces, and give the scores under different levels of variables, as shown in Table 4. Then combined with the expert's policy interpretation document, it can be concluded that the system model is accurate and high in the policy evaluation.

Table 4. PMC score sheet

Txt	X1	X2	X3	X4	X5	X6	X7	X8	X9	Score
HP1	1	1/3	1	2/3	1	2/3	1	1/5	1	7.6
HP2	1/2	1/3	3/5	1/3	1	2/3	1	1/5	1	5.63
JP1	1/4	1/3	4/5	2/3	1	2/3	1/2	0	1	5.22
JP2	3/4	1	4/5	2/3	1	2/3	1	3/5	1	7.48
LP1	1/2	1	4/5	1	1	2/3	1	4/5	1	7.77
LP2	1/4	1/3	2/5	1/3	1	2/3	1/2	0	1	4.48

PMC curved surface image can visualize the calculation results of the PMC model, so this paper draws the PMC curved surface image by calculating the calculation results of the PMC exponential model above. The PMC curved surface image is constructed using the PMC matrix, a three-by-three matrix involving only nine variables. The variables in this paper are exactly nine, which is exactly the number of variables required by the PMC matrix.

The curved surface shows the focus of different provinces at different latitudes. The higher a portion of a curved surface diagram, the higher and more perfect the score of that part, and the lower the part, the lower the score of that part, the content under the dimension needs to be perfected, as shown in Figs. 8, 9, and 10.

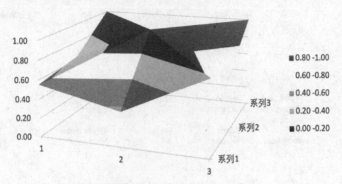

Fig. 8. Heilongjiang PMC curved surface

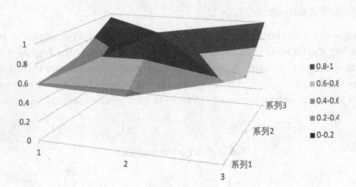

Fig. 9. Jilin PMC curved surface

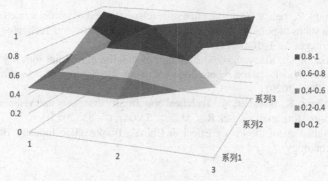

Fig. 10. Liaoning PMC curved surface

4 Conclusion

This paper uses the network crawler to obtain the policy text from the government website, builds a policy evaluation system through A model of the evaluation on text mining, makes the empirical analysis of the use of the policy evaluation system to evaluate the logistics policy, and finally refers to the expert policy interpretation document to prove that this model is effective. The disadvantage of this paper is that the degree of text mining is not deep enough, has not entirely played the role of this article mining technology, but did a relatively shallow level of research. The next step will continue to improve the degree of text mining; word frequency research is not limited to the relationship between segments and segments, will go deep into the relationship between sentences and sentences between the theme words, and further expand text mining results.

References

1. Wei, W., Tiantian, L., Lei, X.: Beijing public creation space support policy tool excavation and quantitative evaluation research. Soft Sci. **32**(9), 56–61 (2018)
2. Howlett, M., Ramesh, M., Perl, A.: Studying Public Policy: Policy Cycles and Policy Subsystems. Oxford University Press, Oxford (2009)
3. McDonnell, L.M., Elmore, R.F.: Getting the job done: alternative policy instruments. Educ. Eval. Policy Anal. **9**(2), 133–152 (1987)
4. Rothwell, R., Zegveld, W.: Industrial innovation and public policy: Preparing for the 1980s and 1990s. Frances Printer, London (1981)
5. Liangcheng, L.: Research on the innovation-driven development strategy policy analysis framework of policy tool dimensions **33**(11), 95–102 (2016)
6. Lixiao, Z.: Research on the theoretical approach to innovative policy assessment - a perspective based on the logical framework of public policy assessment. Sci. Res. **32**(2), 195–202 (2014)
7. Yong'an, Z., Tuo, Z.: Study on quantitative evaluation of policies for mass entrepreneurship and innovation by universal enterprise - Take 10 dual-creation policy intelligence in 2017 as an example
8. Zijun, M., Hong, M.: Comparative analysis of artificial intelligence policy at home and abroad from the perspective of policy tools **39**(04), 74–140 (2020)

9. Wei, W., Yanfa, Z., Lei, X.: Quantification research on the text of artificial intelligence policy in china: the status of policy and the trend of frontiers. Scientific and Technological Progress and Countermeasures 1–10 (2021)

10. Feng, H., Xiaoni, Q., Xiaoyan, W.: Quantitative evaluation of robotics industry policy based on PMC index model - take 8 robotic industry policy intelligence as an example. Intell. J. **39**(01), 121–129 (2020)

11. Yibo, L., Yuhang, K., Shuxuan, W.: Technical opportunity discovery and visual identification based on co-existing analysis. Sci. Res. Manag. **33**(04), 80–85 (2012)

12. Wei, L.: Evaluation of the Policy Effect of China's Photovoltaic Industry Based on Text Mining Technology

MRRVOS: Modular Refinement Referring Video Object Segmentation

Zhijiang Duan[1], Yukuan Sun[2(✉)], and Jianming Wang[1]

[1] School of Computer Science and Technology, TianGong University,
Tianjin 300387, China
{1931085562,wangjianming}@tiangong.edu.cn
[2] Center for Engineering Intership and Training, TianGong University,
Tianjin 300387, China

Abstract. The previous referring video object segmentation method only focuses on the final prediction result. When the user input linguistic query does not match the information in the video, the video still segments error information. We propose a new reference video object segmentation framework. In our model, which we call the Modular Refinement Referring Video Object Segmentation (MRRVOS), when the objects in the referring linguistic query do not match the video frame, stop segmentation and feedback error information. Firstly, given a video clip and a linguistic query. Our method segments the specified object in a video frame automatically. Then our method matches linguistic query with video frames and implements object segmentation on other frames through a recursive three-module model: (1) Referring Video Object Segmentation module: we consider the referring video object segmentation task as a joint problem of referring object segmentation in the image and mask propagation in the video. (2) Image caption module: using recurrent neural networks (RNNs), and deep convolutional neural network (CNN) to encode every frame except the first frame, and a Long Short Term Memory (LSTM) RNN decoder to generate the output caption and put it into the corpus. (3) Semantic dissimilarity module: put all the text results into the corpus and embed vector space, our linguistic query of the input to perform a semantic dissimilarity search. We show that our approach is competitive to the state-of-the-art method.

Keywords: Referring video object segmentation · Semantic similarly · Image caption

1 Introduction

The original intention of the human design computer is to want less error of computer, improve the efficiency of human industrialization. The rapid development of internet videos such as on YouTube and TikTok has brought great

This work was supported by The Tianjin Science and Technology Program (19PT ZWHZ00020).

Y. Gao et al. (Eds.): APWeb-WAIM 2021 Workshops, CCIS 1505, pp. 117–128, 2021.
https://doi.org/10.1007/978-981-16-8143-1_11

challenges to the accurate retrieval of video information. Referring video object segmentation is a common and intuitive method for video segmentation. Given a linguistic query, object instance search aims to segment the images or frames containing the query and locate all objects that appear. Traditional video object segmentation methods [1–3] are mainly based on keyword search [4], predefined keywords, and automatically or manually assigned to the video. However, due to the limitations and unstructured keywords, it is difficult to retrieve various fine-grained content. To address the limitation of the keyword-based approach, more and more researchers are paying attention to video segmentation using natural linguistic texts [5–7] that contain richer and more structured details than keywords.

Fig. 1. The first line of linguistic query and video segmentation match successfully, the second line of linguistic query and video segmentation match failed.

The current dominant approach for cross-modal segmentation is to encode different modalities into a joint embedding space to measure cross-modal similarities, which can be broadly classified into two categories. The first type of works embeds videos and texts as global vectors that encode salient semantic meanings. However, it can be hard for such a compact representation to capture fine-grained semantic details in texts and videos. To avoid losing those details, another type of method employs a sequence of frames and words to represent videos and texts respectively and aligns local components to compute overall similarities.

For referring video object segmentation, the state-of-the-art methods still have the following problems:

1. The segmentation process is not visible and interpretable for users, so human-computer close cooperation is limited.
2. Since "query" is constant for the entire segmentation process, the content of searching is not available to be changed during segmenting.
3. The video still segmentation an error result when the query we enter is mismatched or informally in the video. To solve the above problems, we propose a new reference video object segmentation framework that can be visible to the user.

Figure 1 shows the shortcomings of the existing methods. Enter the linguistic query, we can segment the object we are interested in the video. In the first video, the query of our input is "A woman rides on the horse", the relationship between "Woman" and "Horse" in the entire video has not changed, so the segmentation is successful. But in the second video, we input "Bird standing in the grass". The query is only consistent with the first half of the video, and in the second half of the video, the relationship between the objects in our query does not match the information in the second half of the video. In the previous reference video segmentation methods, we always default to the fixed relationship between the segment targets. However, in daily life, the relationship between the objects we are interested in often changes.

In this work, we propose a new video object segmentation framework, just input a line of the linguistic query, and segment the object we are interested in the video. Segmentation stops and feedback when the relationship between the objects we enter the linguistic query does not match the actual video information. We propose a new task, which is to determine whether the linguistic query matches the video information. We extracted 38 video sequences from DA VIS-2017 and A2D datasets, which are suitable for our task, to form a dataset to determine whether the linguistic query matches the video information. And proposed a new metrics to verify our task.

The contributions of this work are as follows:

1. Our paper proposes text-image and image-text interpretable video object segmentation method.
2. We build a dataset to determine whether the linguistic query matches the video information.
3. When the user input query is not matched with the video content, we stop segmentation and feedback the error information.
4. Semantic dissimilarity module is proposed to analyze the linguistic query user input and image captions for each video frame, and to determine whether there is a mismatch between the linguistic query and the video frame.

2 Related Works

2.1 Modular Networks

In the visual question answering system, the neural module network [8] is introduced. These networks decompose a given question into several components, compute an answer of a given question, and assemble a network dynamically. Modular network has been applied to several other tasks: question answering [9], visual reasoning [10,11], relationship modeling [12], multitask reinforcement learning [13], etc. In the early work [8,9,11], we require to decompose the external language parser, but recent methods [10,12] propose to learn the decomposition end-to-end. We apply this idea to referring video object segmentation, spurn the use of external parser, and taking an end-to-end approach. We find that our soft attention approach achieves better performance than the hard decision predicted by the parser.

2.2 Referring Video Object Segmentation

Traditional VOS [1,2] pays attention to semi-supervised setting and provides binary mask for the first frame of video. Khoreva et al. [5] proposed a linguistic expression to replace mask supervision. In their work, they collected referring expressions of the annotated objects and extend the DAVIS-2017 dataset [14]. They provide two different kinds of annotations from two annotators: first frame annotations are generated only by watching the first frame of the video, whereas full video annotations are generated after watching the whole video sequence. They use the image-based MAttNet [15] model pretrained on RefCOCO to determine the localization of the referred object, and then generate pixel-wise prediction by using DAVIS-2017 to train the segmentation network. Temporal consistency is enforced, so that bounding boxes are coherent across frames, with a post-processing step. Khoreva et al. [5] is the only work previous to ours that focuses on video object segmentation. Related work by Gavrilyuk et al. [16] addresses a similar task by segmenting video objects given a natural language query. They extend the Actor-Action Dataset (A2D) [17] by collecting phrases, but some of them may be ambiguous with respect to the intended referent, as they were not produced with the aim of reference, but description. The authors propose a model with a 3D convolutional encoder and dynamic filters that specialize to localize the referred objects. Wang et al. [18] also leverage 3D convolutional networks, adding cross-attention between the visual and the language encoder. Khoreva et al. [5] Gavrilyuk et al. [16] Wang et al. [18]

2.3 Referring Expression Comprehension and Generation

Since referring expression comprehension and referring expression generation are the counterparts of each other, the researchers usually consider these two tasks simultaneously in many types of research [12,19,20]. [19] proposed to utilize models trained for referring expression comprehension tasks to generate better-referring expressions. [20] explores the role of attributes by incorporating them into both referring expression generation and comprehension. [12] proposed a method that can generate an unambiguous referring expression of a specific object, and which can also comprehend such an expression to infer which object is being described. Now state-of-the-art methods are proposed to decompose a REC problem into modular components (such as subject, relationship and etc.)

and carry out with a modular deep architecture [12,15]. Our work is based on Image-caption [21] and MAttNet [15].

3   Methods

3.1   Overview

In the previous referring video object segmentation task, given a video with N frames and a linguistic query Q, we can get the prediction result of the video for each frame (except for the first frame). However, when the input query doesn't match the information in the video, the video will still segment an error result. We default that the relationship between the segmented targets is always constant, but in daily life experience, the relationship between targets often changes. To solve this problem, we propose a new referring object segmentation framework, just enter a line of the query, we can segmentation the object we are interested in the video, while when the objects in the video do not match the content in the video, stop segmentation and feedback.

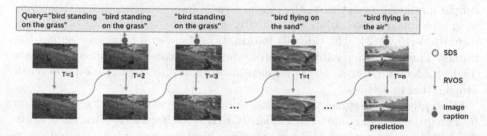

Fig. 2. The flow chart of referring video object segmentation.

3.2   Referring Video Object Segmentation (RVOS) Module

Video Object Segmentation [1,2] (VOS) has been traditionally considered on setups in which the user would manually annotate a frame in the video, and a segmentation system would generate a pixel-wise binary mask for the object in all video frames where it is visible. Our work aims at improving the human-computer interaction by allowing linguistic query as initialization cues, instead of interactive segmentations under the form of a detailed binary mask, bounding box, scribble or point. In particular, we focus on referring query that allow the identification of an individual object or a state in discourse or scene. In Fig. 2, the linguistic query of the first image is "bird standing on the grass".

Linguistic-guided Video Object Segmentation (LVOS) was first addressed by Khoreva et al. [5] and tackled later by Gavrilyuk et al. [16] and Wang et al. [18] Compared to related works on still images, referring query for video objects may be more complex, as they can refer to variations in the properties of the objects, such as a change of location or appearance.

Given a video clip $V = \{x_1\}_{l=1}^{L}$ and input a referring query $r = \{u_t\}_{t=1}^{T}$ to specify an object in X_l, our method calculates the binary object masks * of all the frames in the video $B = \{B_l\}_{l=1}^{l}$.

For any possible referring query $r = \{u_t\}_{t=1}^{T}$, all the u_n forms a dictionary set $D = \{u_i\}_{i=1}^{K}$. We denote all the noun words (each of them specifies an object) as $N = \{u_k\}_{k=1}^{k_1}$, and N is a subset of D. We suppose that any object described by a word in N can be taken as a segmentation instance; otherwise, it is taken as background.

In Video object segmentation research from the previous work, the referring object concept is proposed. Given a video clip and a referring query to specify an object, such methods can specify an object from the video frame, but the effect is limited. In the Fig. 2, "bird standing in the grass", the query can only specify "bird" in the video, ignore the other object "grass". "bird" and "grass" associated with "standing" is equally important in this sentence. Also important, similar examples, such as "a man sitting on the chair".

Given the video and linguistic query Q with N frames, the goal of referring video object segmentation is to predict the binary segmentation mask of the object(s) corresponding to the query Q in N frames. As mentioned earlier, a simple method is to estimate the mask of each frame independently. However, when the image-based solution is directly applied to the referring object segmentation [15, 22–24], the valuable information, temporal coherence across the frames. Therefore, we regard the task of referring video object segmentation as a joint problem of referring object segmentation in an image [15, 22–24] and mask propagation in a video.

For video object segmentation tasks, Collaborative video object segmentation by Foreground-Background Integration (CFBI) [3] effectively takes the feature diversity into account and achieves promising performance. First, in addition to learning feature embedding from foreground pixels, CFBI also considers learning feature embedding from background pixels for collaboration. This learning scheme will encourage the feature embedding from the target object and its corresponding background for comparison, and improve the segmentation results accordingly. Secondly, in the collaboration of foreground and background pixels, further embedding matching is carried out from pixel-level and instance-level. For the pixel-level matching, the module improve the robustness of local matching under various motion rates. For the instance-level matching, the module design an instance-level attention mechanism to enhance the pixel-level matching effectively. Moreover, in order to implicitly aggregate the learned foreground and background information, as well as pixel-level and instance-level information, CFBI use a collaborative ensemble to construct large receptive fields and make accurate predictions.

In Fig. 2, input the linguistic query the prediction mask on the first frame, then cooperate with all the original video frames to output the prediction mask.

3.3 Image Caption Module

In this section, we describe the models that we use for caption generation. Image captions are designed to generate a natural linguistic description of an image. Open-domain captioning is a very challenging task, as it requires a fine-grained understanding of the global and the local entities in an image, as well as their attributes and relationships.

Inspired by the recently introduced example of encoder/decoder using recurrent neural networks (RNNs) for machine translation, and [25] use a deep convolutional neural network (CNN) to encode the input image, and used a Long Short Term Memory (LSTM) [26] RNN decoder to generate the output caption. These systems use back-propagation for end-to-end training and have achieved state-of-the-art results on MSCOCO. More recently in [27], spatial attention mechanism is used to combine visual context on CNN layer, which implies the conditions of the text generated so far. It has been shown and we have qualitatively observed that captioning systems using attention mechanisms lead to better generalization, because these models can compose novel text descriptions based on the recognition of global and local entities that make up the image.

In the process of segmentation, we generate images caption every k frames (except the first frame) and put all the generated caption into the corpus.

3.4 Semantic Dissimilarity (SDS) Module

Given a pre-trained, well-performing cross-encoder, we sample sentence pairs according to a certain sampling strategy and label these using the cross-encoder. These weakly labeled examples will be merged with the strongly labeled dataset. We then train the bi-encoder on this extended training dataset.

In our work, the image caption of the video frame that is started is close to our query, and because the offset of adjacent pixels is not large. Therefore, the score of the initial start is evenly uniform, and there is no big undulation. However, when the status of the video frame or the relationship between the object is changed, it is not matched with the query we entered, and the score we got will have a substantial decline, and we derive feedback from the video does not match the video according to the sudden changes in the gradient.

Our method extracts each h frames, SDS outputs a similarity score g_i, we use the following formula to normalize the output score, so that the score is controlled between $[0, 1]$.

$$G_i = \frac{g_i}{g_1}(i > 2) \tag{1}$$

$$k_i = \frac{G_i - G_{i-1}}{h} = \frac{g_i - g_{i-1}}{g_1 h}(i > 2) \tag{2}$$

Since each video is an independent event, we can't predict the start value of SDS. We extract the first score from the corpus and set it to 1. Then the other scores are divided in the corpus by the first score to get a new score G_i, which will enlarge the score in the corpus. Then we use (2) to calculate the gradient of G_i. After several comparison experiments, when the k value is greater than 3.42×10^{-2}, Our accuracy is the highest.

4   Experiments

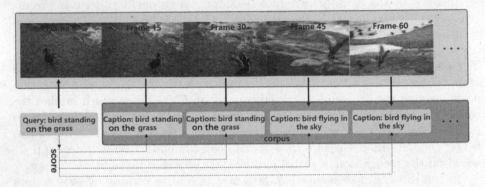

Fig. 3. First line is the current video frame, a linguistic query (purple area) is given. The caption (pink area) of each frame after the first frame is processed, put all the text results into the corpus (yellow area), and compare the similarity in all corpus. (Color figure online)

In the section, we compare our method with the state-of-the-art methods.

4.1   Dataset and Metrics

We extracted 38 video sequences from DAVIS-2017 and A2D datasets, which are suitable for our task, to form a dataset to determine whether the linguistic query matches the video information. It has a total of 2084 labeled frames, a video frame rate is 24 fps, and resolution ratio is 1080p. Our task is to input a referring linguistic query, which can segment the object we are interested in the video, and determine the matching degree between the linguistic query and the video information. When the linguistic query does not match the video information, we stop the segmentation. Since our task is a novel one, this paper defines a new evaluation metric to evaluate the effectiveness of this task: the matching accuracy of linguistic query and video information. This evaluation metric is designed to verify the matching relationship between the linguistic query and each video frame. In our evaluation framework, given the referring linguistic query data G on the first frame of the video, the labeled data B of all video frames, and an output video frame result T. Our evaluation index is to solve a problem: whether the degree of fit or similarity between G and T corresponds to B.

Table 1. Matching accuracy of linguistic query and video information.

Frames (t)	MAttNet	CFBI	Ours
t < 31	0.584	0.555	**0.851**
30 < t < 41	0.631	0.649	**0.874**
40 < t < 51	0.511	0.543	**0.753**
50 < t < 61	0.593	0.613	**0.743**
60 < t < 71	0.663	0.678	**0.864**
70 < t < 81	0.569	0.594	**0.834**
t > 81	0.477	0.523	**0.691**
Mean accurary	0.575	0.619	**0.801**

4.2   Evaluation

Figure 3 illustrates the process of our method. After observing the first frame, given a linguistic query, such as "bird standing on the grass". We perform referring image segmentation on the first frame, then use the segmentation result of the first frame as a benchmark for referring video segmentation. At the same time of video segmentation, we process the image caption of each frame after the first frame, then put all the text results into the corpus and embed them into the vector space. Then use the semantic dissimilarity module to compare the similarity of our initial input linguistic query.

When the input linguistic query is embedded into the same vector space, the score is compared with that of all corpora. For the beginning of the video frame, the offset of adjacent pixels is not large, and the output image caption is very close to our linguistic query, so the score in the semantic dissimilarity module is relatively uniform. However, when the state of the referring object in the video frame or the relationship between multiple referring objects changes, that is, when the video frame information does not match our input linguistic query, the output score of the semantic dissimilarity module will drop significantly. According to Semantic dissimilarity module, we can determine the mismatch between the linguistic query and the video frame. When the object state in the referring linguistic query or the relationship between multiple referring objects does not match the video frame information, we stop segmentation and feedback the error information. Figure 4 shows our comparison with the state-of-the-art methods. Figure 5 shows our quantitative results.

In Table 1, we compare the matching accuracy of our method with the state-of-the-art segmentation method, divide the videos in the dataset into different length sets, and compare our method for multiple rounds on videos of different lengths. Experiments show that with the increase of video frames, the effect of our method is still significant.

T=1   T=20   T=50

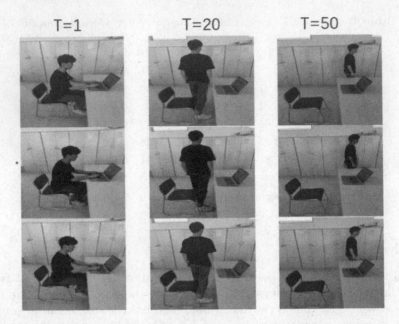

Fig. 4. Input the query: "man sitting in the chair". First line is the original video frame. The second line is the result of the state-of-the-art method. The third line is our method, for the case that the query does not match the video frame, we don't segment and feedback the wrong information to the user.

(a) Dog runs on the ground

(b) Bird standing on the grass

(c) Man walking on the grass

(d) Car on the skateboard

Fig. 5. Qualitative results of our models on DAVIS-2017 dataset.

5   Conclusion

In this paper, We propose a new referring video object segmentation framework, we call Modular Refinement Referring Video Object Segmentation (MRRVOS), when the objects in the referring linguistic query do not match the video frame, stop segmentation and feedback error information. Experiments show that our approach is competitive to the state-of-the-art method.

Acknowledgment. This work was supported by The Tianjin Science and Technology Program (19PTZWHZ00020).

References

1. Perazzi, F., Pont-Tuset, J., McWilliams, B., Van Gool, L., Gross, M., Sorkine-Hornung, A.: A benchmark dataset and evaluation methodology for video object segmentation. In: Proceedings of the IEEE Conference on Computer Vision and Pattern Recognition, pp. 724–732 (2016)
2. Xu, N., et al.: YouTube-VOS: sequence-to-sequence video object segmentation. In: Ferrari, V., Hebert, M., Sminchisescu, C., Weiss, Y. (eds.) ECCV 2018. LNCS, vol. 11209, pp. 603–619. Springer, Cham (2018). https://doi.org/10.1007/978-3-030-01228-1_36
3. Yang, Z., Wei, Y., Yang, Y.: Collaborative video object segmentation by foreground-background integration. In: Vedaldi, A., Bischof, H., Brox, T., Frahm, J.-M. (eds.) ECCV 2020. LNCS, vol. 12350, pp. 332–348. Springer, Cham (2020). https://doi.org/10.1007/978-3-030-58558-7_20
4. Choudhari, K., Bhalla, V.K.: Video search engine optimization using keyword and feature analysis. Procedia Comput. Sci. **58**, 691–697 (2015)
5. Khoreva, A., Rohrbach, A., Schiele, B.: Video object segmentation with language referring expressions. In: Jawahar, C.V., Li, H., Mori, G., Schindler, K. (eds.) ACCV 2018. LNCS, vol. 11364, pp. 123–141. Springer, Cham (2019). https://doi.org/10.1007/978-3-030-20870-7_8
6. Bellver, M., Ventura, C., Silberer, C., Kazakos, I., Torres, J., Giro-i-Nieto, X.: RefVOS: a closer look at referring expressions for video object segmentation. arXiv preprint arXiv:2010.00263 (2020)
7. Seo, S., Lee, J.-Y., Han, B.: URVOS: unified referring video object segmentation network with a large-scale benchmark. In: Vedaldi, A., Bischof, H., Brox, T., Frahm, J.-M. (eds.) ECCV 2020. LNCS, vol. 12360, pp. 208–223. Springer, Cham (2020). https://doi.org/10.1007/978-3-030-58555-6_13
8. Andreas, J., Rohrbach, M., Darrell, T., Klein, D.: Neural module networks. In: Proceedings of the IEEE Conference on Computer Vision and Pattern Recognition, pp. 39–48 (2016)
9. Andreas, J., Rohrbach, M., Darrell, T., Klein, D.: Learning to compose neural networks for question answering. arXiv preprint arXiv:1601.01705 (2016)
10. Hu, R., Andreas, J., Rohrbach, M., Darrell, T., Saenko, K.: Learning to reason: end-to-end module networks for visual question answering. In: Proceedings of the IEEE International Conference on Computer Vision, pp. 804–813 (2017)
11. Johnson, J., et al.: Inferring and executing programs for visual reasoning. In: Proceedings of the IEEE International Conference on Computer Vision, pp. 2989–2998 (2017)

12. Hu, R., Rohrbach, M., Andreas, J., Darrell, T., Saenko, K.: Modeling relationships in referential expressions with compositional modular networks. In: Computer Vision and Pattern Recognition (CVPR), pp. 4418–4427 (2017)
13. Andreas, J., Klein, D., Levine, S.: Modular multitask reinforcement learning with policy sketches. In: International Conference on Machine Learning, pp. 166–175 (2017)
14. Pont-Tuset, J., Perazzi, F., Caelles, S., Arbeláez, P., Sorkine-Hornung, A., Van Gool, L.: The 2017 DAVIS challenge on video object segmentation. arXiv preprint arXiv:1704.00675 (2017)
15. Yu, L., et al.: MAttNet: modular attention network for referring expression comprehension. In: Proceedings of the IEEE Conference on Computer Vision and Pattern Recognition, pp. 1307–1315 (2018)
16. Gavrilyuk, K., Ghodrati, A., Li, Z., Snoek, C.G.: Actor and action video segmentation from a sentence. In: Proceedings of the IEEE Conference on Computer Vision and Pattern Recognition, pp. 5958–5966 (2018)
17. Xu, C., Hsieh, S.H., Xiong, C., Corso, J.J.: Can humans fly? Action understanding with multiple classes of actors. In: Proceedings of the IEEE Conference on Computer Vision and Pattern Recognition, pp. 2264–2273 (2015)
18. Wang, H., Deng, C., Yan, J., Tao, D.: Asymmetric cross-guided attention network for actor and action video segmentation from natural language query. In: Proceedings of the IEEE/CVF International Conference on Computer Vision, pp. 3939–3948 (2019)
19. Luo, R., Shakhnarovich, G.: Comprehension-guided referring expressions. In: Proceedings of the IEEE Conference on Computer Vision and Pattern Recognition, pp. 7102–7111 (2017)
20. Liu, J., Wang, L., Yang, M.H.: Referring expression generation and comprehension via attributes. In: IEEE International Conference on Computer Vision (ICCV), pp. 4866–4874 (2017)
21. Luo, R., Price, B., Cohen, S., Shakhnarovich, G.: Discriminability objective for training descriptive captions. In: Proceedings of the IEEE Conference on Computer Vision and Pattern Recognition, pp. 6964–6974 (2018)
22. Li, R., et al.: Referring image segmentation via recurrent refinement networks. In: Proceedings of the IEEE Conference on Computer Vision and Pattern Recognition, pp. 5745–5753 (2018)
23. Margffoy-Tuay, E., Pérez, J.C., Botero, E., Arbeláez, P.: Dynamic multimodal instance segmentation guided by natural language queries. In: Ferrari, V., Hebert, M., Sminchisescu, C., Weiss, Y. (eds.) ECCV 2018. LNCS, vol. 11215, pp. 656–672. Springer, Cham (2018). https://doi.org/10.1007/978-3-030-01252-6_39
24. Ye, L., Rochan, M., Liu, Z., Wang, Y.: Cross-modal self-attention network for referring image segmentation. In: Proceedings of the IEEE/CVF Conference on Computer Vision and Pattern Recognition, pp. 10502–10511 (2019)
25. Karpathy, A., Fei-Fei, L.: Deep visual-semantic alignments for generating image descriptions. In: Proceedings of the IEEE Conference on Computer Vision and Pattern Recognition, pp. 3128–3137 (2015)
26. Hochreiter, S., Schmidhuber, J.: Long short-term memory. Neural Comput. 9, 1735–1780 (1997)
27. Xu, K., et al.: Show, attend and tell: neural image caption generation with visual attention. In: International Conference on Machine Learning, pp. 2048–2057 (2015)

Author Index

Printed in the United States
by Baker & Taylor Publisher Services